"This a 4-year college class and 10-year internship rolle

—Mark Seigle, Industry Executive

"This is an absolute must-read and a necessary resource for any media executive, educator, or student wanting to have a comprehensive understanding of the myriad of descriptions, issues, and impacts on a variety of technical elements and management perspectives focusing on the workflow involved in the production, post-production, distribution, asset management, and archiving of media content. What Chris Lennon and Clyde Smith (two of the most knowledgeable experts in the world on this topic) have done here in their book is to put all of these pieces of the media workflow puzzle together in one place, providing a most intriguing assembly of topics and knowledge that covers this rapidly evolving terrain with tremendous insight and clarity."

—Dr. Corey P. Carbonara, Professor, Department of
Film and Digital Media, Baylor University

"Media workflow is a complex and nuanced area. Chris Lennon and Clyde Smith have compiled a deep look at the component parts of this area and give the reader tremendous context into how the pieces fit together with brilliant business context."

—Harold S. Geller, Executive Director, Ad-ID LLC.

"A great overview of workflows in the media industry. Gives a great foundation about how and why TV and film technologies got to this point, how they operate, where they come together, and it also provides a good look at where the industry may be headed in the future. This is a great read for college students, people entering the media business, and for those people like me who are working in the entertainment business."

—Tom Kline, Sr. Director of Program Standards, a major US network

"Nothing else in this world compares to electronic media workflow. The machinery, tools, and processes are constantly evolving and the capabilities constantly increasing. This book follows the evolution from those simple signal flows to virtualization and abstraction. There are plenty of takeaways. Plenty of 'so this is why we do it this way' moments. I think fortunately, for our own enjoyment of electronic media entertainment and education, we all fail to properly appreciate workflow as a result of overfamiliarity. At least for the time one is reading this book we can admire the elegance and enchanting mechanisms that step-after-step make up what we all, even us workflow wonks, take for granted. It's a good read. It makes you stop and think 'What if we did this differently?' The creative part of electronic media isn't just the content – this book captures that."

—Fred Baumgartner, Director NextGen TV Implementation,
ONEMedia 3.0 / Sinclair Broadcasting

"A must read for both experts and non-experts in both cinema and TV technology."

—Pierre Maillat, Canal+

The Media Workflow Puzzle

This edited collection brings together a team of top industry experts to provide a comprehensive look at the entire media workflow from start to finish.

The Media Workflow Puzzle gives readers an in-depth overview of the workflow process,
from production to distribution to archiving. Pulling from the expertise of twenty contributing authors and editors, the book covers topics including content production, postproduction systems, media asset management, content distribution, and archiving and preservation,
offering the reader an understanding of all the various elements and processes that go into
the media workflow ecosystem. It concludes with an exploration of the possibilities for the
future of media workflows and the new opportunities it may bring.

Professionals and students alike looking to understand how to manage media content for
its entire lifecycle will find this an invaluable resource.

Chris Lennon is President and CEO of MediAnswers, a trusted advisor to the cream of the
crop of media companies worldwide. He has over 30 years of experience in the media business, leading large projects on every continent except for Antarctica. He has held a variety
of leadership positions in the Society of Motion Picture and Television Engineers (SMPTE),
including Standards Director, Local Section Manager, Technology Committee Chair, and
others. He is an SMPTE Fellow and recipient of SMPTE's 2008 Citation. He is known as a
thought leader, with a reputation built on not only working with the world's leading media
organizations, but his leadership of several leading-edge industry efforts in many diverse
areas, like BXF, NABA/DPP Content Delivery Specs, PMCP, 3DTV, OBID Audience Measurement, Multilink SDI using CWDM, microservices in media, and many more.

He also serves as Executive Director of the Open Services Alliance, a global industry organization focused on interoperability among microservice-oriented media systems.

In his spare time, he is a championship winning race car driver, author, and high-
performance driving coach, and runs Winding Road Adventures.

For more details on MediAnswers, its clients, and Chris's team, please visit www.median-
swers.tv. For more details on the Open Services Alliance, please visit www.openservicesal-
liance.com.

Clyde Smith is a Retired Former Senior Vice President, Advanced Technology for FOX Network Engineering and Operations, where he supported Broadcast and Cable Networks operating groups in addressing their challenges with technologies, standards, and regulations.

Previously, he was SVP of Global Broadcast Technology and Standards for Turner Broadcasting System, Inc. In this capacity, Smith was responsible for the strategic development

and planning of new technology in addition to the operational transition of media from production to broadcast to air. Smith's broad knowledge of operational and technical systems and hands-on experience developing processes for integrating operational facilities from the ground up were vital to the Turner Entertainment Group's continuing expansion.

Prior positions include SVP of Broadcast Engineering Research and Development at Turner and SVP & CTO at Speer Communications where he managed operations of one of the nation's first all-digital facilities.

He worked for 8 years in communications design and development engineering at the Kennedy Space Center where he supported 48 shuttle missions, three interplanetary probes, and numerous Department of Defense initiatives.

Smith supported initiatives that were recognized by The Computer World Honors program with the 2005, 21st Century Achievement Award and an Emmy Award for Pioneering Efforts in the Development of Automated, Server-Based Closed Captioning Systems.

Smith is a frequent speaker and honored guest at meetings for SMPTE, NAB, and SBE. He served 15 years as an SMPTE Governor; he also served as a Standards Chairman and Secretary/Treasurer of the SMPTE. He also was program chair for four SMPTE advance-imaging conferences. He has often presented and published research and technology papers that he has authored or co-authored in industry magazines and at industry conferences including CES, NAB, Hollywood Professional Alliance Tech Retreats, Storage Networking World, Storage Visions, VidTrans, IBC, SMPTE, SPIE, UFVA, and SBE.

He is an SMPTE Fellow and recipient of the SMPTE Progress Medal, SMPTE David Sarnoff Medal Award, and the SMPTE Outstanding Service Award as well as The Broadcasting and Cable Technology Leadership Award and the Storage Visions' Storage Industry Service Award. He is a recipient of the North American Broadcasters Association International Achievement Award and an Honorary Member of the International Association of Broadcast Manufacturers (IABM).

The Media Workflow Puzzle

How It All Fits Together

Edited by Chris Lennon and Clyde Smith

R Routledge
Taylor & Francis Group

NEW YORK AND LONDON

First published 2021
by Routledge
52 Vanderbilt Avenue, New York, NY 10017

and by Routledge
2 Park Square, Milton Park, Abingdon, Oxon, OX14 4RN

Routledge is an imprint of the Taylor & Francis Group, an informa business

© 2021 Taylor & Francis

Library of Congress Cataloging-in-Publication Data
Names: Lennon, Chris (Technologist), author. | Smith, Clyde (Broadcast engineer), editor.
Title: The media workflow puzzle: how it all fits together / edited by Chris Lennon and Clyde Smith.
Description: London; New York: Routledge, 2021. | Includes bibliographical references and index. |
Identifiers: LCCN 2020043547 | ISBN 9780815392897 (hardback) | ISBN 9780815392903 (paperback) | ISBN 9781351189552 (ebook)
Subjects: LCSH: Mass media—Management. | Digital media—Management. | Multimedia systems-—Management. | Information resources management.
Classification: LCC P96.M34 M435 2021 | DDC 302.23/068—dc23
LC record available at https://lccn.loc.gov/2020043547

ISBN: 978-0-815-39289-7 (hbk)
ISBN: 978-0-815-39290-3 (pbk)
ISBN: 978-1-351-18955-2 (ebk)

Typeset in Berling and Futura
by codeMantra

Contents

Contributors

EDITORS

Chris Lennon is President and CEO of MediAnswers, a trusted advisor to the cream of the crop of media companies worldwide. He has over 30 years of experience in the media business, leading large projects on every continent except for Antarctica. He has held a variety of leadership positions in the Society of Motion Picture and Television Engineers (SMPTE), including Standards Director, Local Section Manager, Technology Committee Chair, and others. He is an SMPTE Fellow and recipient of SMPTE's 2008 Citation. He is known as a thought leader, with a reputation built on not only working with the world's leading media organizations, but his leadership of several leading-edge industry efforts in many diverse areas, like BXF, NABA/DPP Content Delivery Specs, PMCP, 3DTV, OBID Audience Measurement, Multilink SDI using CWDM, microservices in media, and many more.

He also serves as Executive Director of the Open Services Alliance, a global industry organization focused on interoperability among microservice-oriented media systems.

In his spare time, he is a championship winning race car driver, author, and high-performance driving coach, and runs Winding Road Adventures.

For more details on MediAnswers, its clients, and Chris's team, please visit www.medianswers.tv. For more details on the Open Services Alliance, please visit www.openservicesalliance.com.

Clyde Smith is a Retired Former Senior Vice President, Advanced Technology for FOX Network Engineering and Operations, where he supported Broadcast and Cable Networks operating groups in addressing their challenges with technologies, standards, and regulations.

Previously, he was SVP of Global Broadcast Technology and Standards for Turner Broadcasting System, Inc. In this capacity, Smith was responsible for the strategic development and planning of new technology in addition to the operational transition of media from production to broadcast to air. Smith's broad knowledge of operational and technical systems and hands-on experience developing processes for integrating operational facilities from the ground up were vital to the Turner Entertainment Group's continuing expansion.

Prior positions include SVP of Broadcast Engineering Research and Development at Turner and SVP & CTO at Speer Communications where he managed operations of one of the nation's first all-digital facilities.

He worked for eight years in communications design and development engineering at the Kennedy Space Center, where he supported 48 shuttle missions, 3 interplanetary probes, and numerous Department of Defense initiatives.

Smith supported initiatives that were recognized by The Computer World Honors program with the 2005, Twenty-First Century Achievement Award and an Emmy Award for Pioneering Efforts in the Development of Automated, Server-Based Closed Captioning Systems.

Smith is a frequent speaker and honored guest at meetings for SMPTE, NAB, and SBE. He served 15 years as an SMPTE Governor; he also served as a Standards Chairman and Secretary/Treasurer of the SMPTE. He also was program chair for four SMPTE advance-imaging conferences. He has often presented and published research and technology papers that he has authored or co-authored in industry magazines and at industry conferences including CES, NAB, Hollywood Professional Alliance Tech Retreats, Storage Networking World, Storage Visions, VidTrans, IBC, SMPTE, SPIE, UFVA, and SBE.

He is an SMPTE Fellow and recipient of the SMPTE Progress Medal, SMPTE David Sarnoff Medal Award, and the SMPTE outstanding service award as well as The Broadcasting and Cable Technology Leadership award and the Storage Visions' Storage Industry Service Award. He is a recipient of the North American Broadcasters Association International Achievement Award and an Honorary Member of the International Association of Broadcast Manufacturers (IABM).

SECTION EDITORS

Brian Campanotti is recognized as a successful entrepreneur, thought leader, published author and innovator in the area of large-scale unstructured data storage, archive and preservation. He is the founder of Cloudfirst.io, an innovative end-to-end technology and services provider focused on digital archive transformation and modernization, helping global content owners, producers and custodians develop and execute on their long-term strategies for massive-scale digital content archive, protection and preservation.

Prior to founding Cloudfirst.io, Mr. Campanotti was CTO at Front Porch Digital and instrumental in their emergence as the global leader in the digital archive space. During his tenure, he helped guide the company through several successful transactions including its acquisition by Oracle in 2014. Prior to that, he founded Masstech, another leading provider of digital archive solutions, and began his career as a Project Engineer at the Canadian Broadcasting Corporation (CBC) in Toronto, Canada.

Mr. Campanotti and his team have won three Emmy® Awards for pivotal innovation in digital archive technologies, long-term digital preservation and large-scale digital video implementations. He was one of the primary inventors of the SMPTE and ISO/IEC Archive eXchange Format (AXF) Standard and holds a degree in Electrical Engineering from the University of Toronto.

Arjun Ramamurthy was most recently the Senior Vice President of Technology at Twentieth Century Fox/Disney. In that capacity, he was responsible for setting technology direction for Motion Picture and TV Production, Post-production Digital Content Processing, and downstream distribution and Digital Archiving. In addition, he was responsible for outlining and defining the next generation workflow as well as introduction of new technology and toolsets into the production.

He has over 25 years of experience in the industry, and was previously with Deluxe's EFILM facility, and prior to that, with Warner Bros. Technical Operations and Feature Animation. He is an active member of SMPTE and IEEE, and has contributed on a variety of technical committees and standards. He holds several patents in the area of Digital Image processing and Media Post Production.

He is a member of the Academy of Motion Picture Arts and Sciences and a fellow of Society of Motion Picture and Television Engineers.

Glenn Reitmeier is widely recognized as a technology visionary and pioneer in the television industry. Throughout his career, he has been a leader in establishing revolutionary new digital standards, including the SMPTE Component Digital Video and SDI standards, the Grand Alliance digital HDTV system that became ATSC 1.0, and the new ATSC 3.0 standard. Now an independent consultant, Reitmeier recently retired from 17 years at NBC Universal as SVP, Technology Standards and Policy, where he contributed to industry technical standards and to the technical aspects of the company's government policy positions and commercial agreements. Previously, he spent 25 years in digital video research at RCA/Sarnoff Laboratories. He has served the industry as a Board member of ATSC, NABA, and OATC and has been Chairman of both ATSC and OATC. He is an SMPTE Fellow and a recipient of the Progress Medal and the Signal Processing Medal. He is also an inaugural member of the CTA's Academy of Digital Television Pioneers and a recipient of the NAB Television Engineering Award. He holds over 60 patents and is recognized in the New Jersey Inventors Hall of Fame.

Jay Veloso Batista is a published author and editor. Batista is the Chief Revenue Officer for a SaaS platform supplying innovative technology to the Galleries and Museums Industry. Over the past 30+ years, Batista has held executive positions in media vendor companies including traffic software, rights software, encode/transcode tools, and complete satellite and broadcast transmission systems.

CONTRIBUTING AUTHORS

François Abbe has spent 20+ years in professional video, starting in England as an R&D engineer, then product marketing manager in technology for TV. In 2005, he became an architect and independent consultant, then launched the start-up MESCLADO where he combined innovation, strategy, and HR. His references include media (Canal+, Orange, France Télévisions, ARTE, RTBF), service providers, and sports federations. Thanks to this experience, Abbe now offers an efficient and innovative approach to help executives strengthen and differentiate their communication.

Bryce Alden has spent nearly 15 years in the media entertainment industry focusing on Digital Cinema mastering and distribution. At Deluxe Entertainment Service, he has helped evolve the company from its historic film focus to a highly diversified post-production digital service provider. In his time with Deluxe, Alden has designed and built out digital cinema mastering systems, projection systems and screening rooms, hard-drive duplication systems, and associated workflows. As Vice President of Digital Cinema and Security, he now oversees the operations of Deluxe and DCDC's satellite distribution networks, a system for which he was a key designer. He holds a BS in physics from the University of Southern California.

Wendy Aylsworth (an SMPTE Fellow and past president) has spent over 30 years in entertainment technology, bringing emerging technologies into production and distribution usage, and continues to provide technical consulting and strategic board guidance. She was a key leader in the transition and standardization of digital cinema that has become the mainstay of the industry. She led the digital transitions of the entire animation processes at Disney and Warner Bros., as seen in such hits as Lion King and Space Jam, and new technologies such as high frame rate, as seen in the Hobbit trilogy. Aylsworth is involved in many industry organizations to share knowledge and grow the next generation of entertainment professionals. In recognition of her work, Aylsworth has been honored with many awards, including the Charles F. Jenkins Lifetime Achievement Emmy Award from the Academy of Television Arts & Sciences and the Bob Lambert Technology Leadership Award from the Entertainment Technology Center at the University of Southern California, and is a Lifetime Fellow of the Society of Motion Picture and Television Engineers. Aylsworth holds a BSCE from the University of Michigan and an MS/MBA (Beta Gamma Sigma) from USC.

Annie Chang is the Vice President, Creative Technologies for Universal Pictures, responsible for developing strategies and designing innovative next-generation workflows across film and emerging immersive media experiences. Prior to joining Universal Pictures, Chang was the VP, Technology for Marvel Studios, and held various roles during more than a decade tenure at The Walt Disney Studios. Chang is the AMPAS ACES Project Chair and has been a Co-Chair of the 10E Essence Technology and Chair of the Interoperable Master Format (IMF) Working Group at SMPTE. She is an SMPTE Fellow and is a recipient of the SMPTE Workflow Systems Medal Award, Advanced Imaging Society's Distinguished Leadership award, and StudioDaily's 2018 Exceptional Women in Production and Post. In 2020, The Hollywood Reporter named Chang on their list of top "Hollywood Innovators."

Julián Fernández-Campón has been working in computer-related companies since 1997 when he started with Telecommunication giant Alcatel as software developer. In 2000 he moved to the university research group that would become Tedial, where he was involved with software development and system analysis for the first versions of Tedial products.

From 2004 onwards Fernández-Campón moved into project management. Since then, he has been working in system dimensioning, workflows definition, and broadcast consultancy, and has been involved in most of Tedial's broadcast projects worldwide. He is now Chief Technology Officer at Tedial leading the transition over the last years to meet the new Challenges in the market with the adoption of the latest technologies in the Tedial Products, Solutions, and Services and the enterprise transformation to DevOps.

He has a degree in Computer Science Engineering at the Higher Technical School of Computer Engineering, University of Málaga, specializing in Telecommunications and Robotics.

Frans de Jong holds a Master's degree in Information Theory from Delft Technical University. He started his career in broadcasting as a radio engineer and subsequently worked as a video editor and systems architect for several broadcasting organizations. In 2003 Frans joined the European Broadcasting Union (EBU) in Geneva (Switzerland). He is a Senior Engineer at the Technology & Innovation Department, where he coordinates project groups on production technology topics, such as (U)HDTV, Loudness, Subtitling, Quality Control, and Cloud Production.

Brendan Kehoe is the President of Effective Media Services, an Entertainment Software and Consulting Company specializing in On Air Promotion. Mr. Kehoe has held senior management roles at Fox, Warner Bros., and CBS spanning a 40-year career in Entertainment.

Al Kovalick has specialized in professional networked media and infrastructure for the past 25 years. He started his career at Hewlett-Packard, where he became the principal architect of HP's first Video-on-Demand server. Following HP in 1999, he became the CTO at Pinnacle Systems. In 2004 Kovalick moved to Avid Technology as a Corporate Fellow. In 2012, he founded Media Systems Consulting in Silicon Valley where his clients include the most respected broadcasters and media vendors in the Americas, Asia, and Europe.

He is an active speaker and educator and has presented more than 50 papers at industry conferences. He has authored 21 peer-reviewed articles in the *SMPTE Motion Imaging Journal* and holds 13 U.S. patents. Al wrote the first book of its kind, *Video Systems in an IT Environment: The Basics of Professional Networked Media and File-based Workflows* (2009, 2nd ed.). He has a BSEE degree from San Jose State University and an MSEE degree from the University of California at Berkeley. He is an SMPTE Life Fellow and a recipient of the SMPTE David Sarnoff Medal. Al created UFTmachine.com, an educational website for physics enthusiasts.

Shawn Maynard is the Senior Vice President and General Manager of Florical Systems. Prior to Florical He was the Director of Operations for NBC Universal Local Media Group managing the SouthEast Hub. He also serves on IABM's Americas Council and Secretary/Treasurer of SMPTE's Florida Chapter.

Stan Moote is the CTO for IABM, the international trade association for suppliers of broadcast and media technology.

He began his television career in 1977 interning as a plant engineer for CFTO-TV in Toronto, Canada, during the co-op component of his engineering degree from University of Waterloo.

In 1980, Mr. Moote co-founded Digi-tel and was responsible for the design and development of various innovative digital video products, before bringing his many talents to Leitch in 1984. Mr. Moote was involved in the SMPTE Digital Video Standards Committee meetings, creating CCIR-601, and continued his standardization work on video transport by being on the VSF board of directors, 2001 to 2004. He is an active member of the NATAS Technical EMMY Committee.

With his continued involvement with SMPTE, in 2015 he received the prestigious SMPTE Digital Processing Medal Award.

While holding Vice President and CTO positions at Leitch and Harris, Mr. Moote focused on workflow solutions, new technology, standardization, and interoperability on a global basis. He developed several patents including scrambling systems, data monitoring, multiviewer, router processors, and IPTV systems.

Mr. Moote is an accomplished presenter and journalist. He regularly presents at IBC, NAB, Broadcast Asia, and BIRTV, and is much sought-after for keynotes, panels, and to arrange/moderate various conference sessions. This is all due to his clear understanding of new technology trends, how they interact, and seeing practical applications that both engineers and business managers value worldwide. You can read many of his publications at: ca.linkedin.com/in/stanmoote.

Andy Quested started as a BBC Technical Assistant in 1978 becoming a video-tape editor in 1985 where he worked on many comedy and children's and documentary series. In 1998 Andy moved to a new technology department working on the BBC's first HD programs including Planet Earth I and the first UHD program, Planet Earth II.

He is technical lead of the UK's Digital Production Partnerships AS-11 format and leads the EBU Production Strategic Group looking at all areas of content production. He is an active member of SMPTE becoming a Fellow in 2014 for work relating to standards.

Quested is currently the chair of ITU-R Working Party 6C, where he initiated new areas of study on Advanced Immersive Audio-Visual Systems and Artificial Intelligence in content production and international program exchange.

Karyn Reid is VP, Broadcast Systems at Fox Corp. Reid has been a Product Manager at broadcast traffic/automation, and program management software vendors and has been involved in the SMPTE BXF schema committee since its inception.

Jeff Stansfield is the President and CEO of Advantage Video Systems, a leading technology provider to the broadcast, motion picture, television, and motion graphics industries. He has served as a general board member, treasurer, and secretary of SMPTE (Society of Motion Picture and Television Engineers). He supports many other industry organizations including the Creative Pro User Group Network, the Digital Cinema Society, the Hollywood Post Alliance, the Society of Television Engineers, and the National Association of Broadcasters. His more than 34 years of experience includes the construction of broadcast facilities as well as technology services for production and postproduction facilities, special effects businesses, and motion graphics businesses.

Stansfield newest venture is in the collegiate esports industry. His new company Esports Circus is the ultimate Esports, VR & Robotics mobile venue serving the collegiate and amateur markets to compete, develop advancement, and to have fun! EsportsCircus also helps higher education schools create and fund their Esports teams. See more at https://esports-circus.com/.

Acknowledgments

A very special thank you to Jay Veloso Batista for yeoman's work on editing this book, and to Kathleen Lennon for keeping it all on track. For the excellent cover design, we have the talented Greg Lennon to thank.

The editors would like to thank the following people and organizations for their support and assistance in the considerable undertaking that was the assembly of this book.

PEOPLE
Section Editors

Jay Veloso Batista
Brian Campanotti
Arjun Ramamurthy

Chris Lennon
Clyde Smith
Glenn Reitmeier

Contributing Authors

Francois Abbe
Bryce Alden
Wendy Aylsworth
Brian Campanotti
Annie Chang
Frans De Jong
Julian Fernández-Campón

Brendan Kehoe
Al Kovalick
Shawn Maynard
Stan Moote
Karyn Reid
Andy Quested
Jeff Stansfield

Organizations

Academy of Motion Picture Arts and Sciences
DCI – http://www.dcimovies.com/
ITU-R Recommendations – https://www.itu.int/pub/r-rec (BS – Sound, BT – Television)
ITU-R Reports – https://www.itu.int/pub/r-rep (BS – Sound, BT – Television)
International Association of Broadcast Manufacturers
National Association of Broadcasters
Society of Motion Picture and Television Engineers
The VES Handbook of Visual Effects Industry Standard VFX Practices and Procedures

CHAPTER 1

Introduction

Section Editors: Chris Lennon & Clyde Smith

Welcome to the wonderful world of media workflows! Many take the process of getting audiovisual media from initial production all the way to distribution to the viewer largely for granted. In many cases, it just seems to happen as if by magic. Things have evolved a great deal from the early days of shooting a production on film, editing using human hands and razor blades, creating physical prints, and shipping those to theaters or later to television networks so that viewers could enjoy the finished product.

Today, as a result of over a century of evolution and continual improvement, we have amazingly capable (and complex!) systems and processes in place. The fact that these all work together in a way that seems like "magic" to those not involved in the day-to-day creation of audiovisual content is a testament to the hard work of many over the years to get where we are today.

Plenty of publications exist that can help you to dig deep on various technologies and approaches in use in the entire world of media workflow. This book makes no attempt to replicate or replace any of those.

This book is different. Its intent is to provide a very high-level overview of what's involved in the process, start to finish, of producing audiovisual content.

Who should read this? If you're new to the media business or a student, this should serve as a good place to start. If you're in the business but have a limited view into a relatively small portion of the process, this book should also help you. If you're an executive, not needing to know the details of exactly how things happen technically but need a solid grounding in the process as a whole, you should also find this book helpful.

As you progress through this book and find that you'd like to dig deeper into a specific area, we highly recommend you search for some of the excellent publications available that dig into the technical details you seek. In many cases, the section editors and authors of this book are terrific resources themselves and would be only too happy to point you in the right direction.

We hope you find this book a reliable high-level resource. Our aim is to provide you with a solid appreciation for all the various elements and processes that go into the media workflow ecosystem.

CHAPTER 2

Overview

Section Editors: Chris Lennon & Clyde Smith

Like all industries that survive in our changing world, the ability to transform and adapt has been key in the media business.

When one considers where things started in the movie industry in the early twentieth century and where they now are just over a century later, the rate of change boggles the mind. The media business has been in a constant state of evolution ever since it began. In fact, it could be argued that this evolution has been interspersed with several revolutions along the way.

REVOLUTION #1 – TELEVISION ENTERS THE SCENE

Consider the sea changes that occurred when television first became a popular medium. For the first several decades, the world of audiovisual content primarily revolved around Hollywood and its studios. The number of players was limited and geographically concentrated. This meant that establishing standardized workflows was relatively easy. Techniques were developed and implemented with relative ease among this targeted group. Production, Post-Production, Managing and Archiving Assets, and Distribution was well-defined, and matured to a great degree.

When television entered the scene, it had several important impacts. Production became quicker. Networks now cranked out dozens of episodes per year. Live events had their own workflows by necessity. Managing large quantities of programming and ad content became a reality.

REVOLUTION #2 – WE GO DIGITAL AND FILES REPLACE PHYSICAL MEDIA

The shift to digital and file-based workflows in both Motion Pictures and Television changed everything again. Some think of the switch to digital in a transmission context but although that was a big change, the impacts all the way upstream to production of digital workflows made this perhaps the biggest revolution ever in our business.

It's easier to make a list of what didn't rather than what did change during this revolution. Very little was left untouched by the shift from analog to digital. Every piece of equipment involved in media workflow in the analog world suddenly became obsolete, but this was not just a "lift and shift" to digital equipment. The way things were was transformed and replaced entirely. The replacement of physical media (film, tape, etc.) with digital files opened up entirely new approaches to media workflows. This didn't happen overnight, as this shift was so dramatic it took time for those involved to fully appreciate the opportunities for dramatic alterations in workflows that could be realized through the new technology.

Cost and complexity involved in producing audiovisual content plummeted. Specific equipment that could only be afforded by the elite few was now democratized into "apps" on computers, available for very little cost. Virtually anyone who wanted to could now compete with "the big guys," and produce content that in many cases was competitive in every way with so-called "professionally produced content."

REVOLUTION #3 – MULTIPLATFORM DISTRIBUTION

Things were upset again when distribution expanded to a plethora of new methods, including cable, satellite, Internet, mobile, social media, and the list goes on. Again, although on the surface this would seem to be purely a revolution in the distribution realm, that's not really true. Whereas the Television Revolution had impacts far upstream of distribution, Multiplatform Distribution upset things even more. Workflows and practices all the way up to Production had to be re-examined again. Content being produced must now be suitable for viewing on screens from a few inches to wall sized. Long-form content that worked well for decades wasn't so well-suited to mobile viewing or to social media viewing, where attention spans are counted in second, not minutes or hours.

Competition went from a few studios and a few networks to a virtually limitless number of media outlets offering content to the world via the Internet.

REVOLUTION #4 – THINGS GET CLOUDY

Today, we see further revolution taking place with the onset of Machine Learning/Artificial Intelligence and its application to media, alongside the movement toward cloud-based processes and storage.

The onset of "The Cloud" has further moved media workflow into a world of lower cost and high efficiency. While earlier revolutions moved things more and more away from specialized and expensive equipment, this revolution made any type of equipment almost an afterthought. Producers of content can now simply invoke processes online without ever having to purchase any equipment at all. They can pay by the minute for hardware and software resources, allowing them to scale up and down on demand.

One could say this further democratized the media business, removing some of the last barriers to smaller players that want to get involved.

All of this has meant that processes in some ways became simpler, while at the same time becoming incredibly complex on another level. For the user, it's hard to argue that things have not become greatly simplified. However, in order to make all of this work seamlessly and efficiently, operations "under the covers" have become incredibly complex. All of the diverse systems involved in media workflows don't just work together by default. A large amount of care and attention to the small details must be taken to ensure that this complex ecosystem operates as expected under the increasingly complex demands put on it every day.

In the pages that follow, we will walk you through the process, in a logical sequence from start to finish.

We will begin with Production, addressing with how audio and visual essence is captured.

We will then move onto Post-Production where all of the various elements come together to form a finished product, no small task.

Next, we will talk about managing those assets and their associated workflows which has become increasingly challenging with the sheer volume and diversity of the versions of assets that exist.

Distribution naturally follows. As we mentioned earlier, things have come a long way from distribution consisting of films being shipped to theaters. It takes so many different forms today that it's not a simple matter.

Last, but certainly not least, is Archiving and Preservation. As the cost and value of content increases, it is even more important than ever to properly archive and preserve it for future use. The "long tail" of monetizing content as long as possible after its initial distribution can be a make or break factor in the financial viability of a production. And while one might think that the move to digital has made this all much simpler, it has actually complicated the process a great deal.

We will conclude with a look into our crystal ball at what might be in the future of media workflow. What is likely to be the next (r)evolution we need to pay attention to? What are the likely impacts and opportunities?

CHAPTER 3

Production

Section Editor: Jay Veloso Batista

Production of media is the beginning of our workflow chain. Most often, compelling content is produced through a collaboration of talented individuals and technological tools. To provide you a basis in understanding, we need to start with the origination and capture of the media itself, along with its metadata. This section covers both motion picture and video production together because, while for many years they were separate processes with film dominating the cinema productions, modern tools supply quality media and have relegated film to specialty projects.

CAMERAS

Capturing images began when scientists discovered the light sensitive properties of certain chemicals and began to experiment with substrates and supports, leading to the forerunner of the modern camera and the initial blossoming of tintypes during the mid-nineteenth century. By the twentieth century, innovations had led to motion capture – a video camera is a camera used for electronic motion picture acquisition initially developed for the television industry but now common in all applications.

The earliest video cameras based on the mechanical Nipkow disk were designed by John L. Baird and used in experimental broadcasts from 1918 to the 1930s. All-electronic designs based on the video camera tube, such as Vladimir Zworykin's Iconoscope and Philo Farnsworth's image dissector, replaced the Baird system by the 1930s. These remained in wide use until the 1980s, when technological break-throughs introduced solid-state image sensors such as CCDs and CMOS active pixel sensors into digital camera systems, completely eliminating common tube technologies problems such as "image burn-in" where an overly bright light or a stationary picture would imprint on the tube. For the first time, these developments made digital video workflows practical. Around the world, digital television gave a boost to the manufacture of digital video cameras and by the 2010s, most video cameras were digital for professional and consumer applications.

With digital video capture an affordable technology, the distinction between professional video cameras and movie cameras disappeared. Today the mid-range cameras exclusively used for television and other work are termed professional video cameras.

Creating content with video cameras is dedicated to two core industrial applications. The first, a reflection of the early days of broadcasting, is live event production, where the camera provides the source of real-time images directly to a screen for immediate viewing. While a few production systems still serve live television, especially sport event production, most live camera connections are dedicated to security, police, military, and industrial situations where monitoring is required. In the second application, the images are recorded to a storage device for archiving or further processing. For many years, videotape was the primary format used for this recording, although gradually it was replaced by optical disc, hard disk, and then flash memory. Today, recorded video is the basis of television and movie production, and more often surveillance tasks where unattended records of a situation are required for post event analysis.

Jeff Stansfield of Advantage Video in Los Angeles provides this history lesson on the evolution of camera and associated technology.

For us to really understand the Audio and Video acquisition formats, we should spend a little time first looking into where we came from and understand the history of media acquisition. As we look at the past, we can see the battles and choices that led us to where we are today. Choices really started back in 1982 when Sony developed a professional videocassette product called the Betacam then the Betacam SP and high-end digital recording systems like the Digital Betacam. This was an important step on the path to where we are now for three reasons.

Firstly, the videocassettes all use the same shape with only very slight variations, meaning vaults and other storage facilities do not have to be changed when using a new format. This saved the studios and production companies tons of money in storage fees. It also let a production go right from the set to post without going through any developing or film processing.

Secondly, this started us to bring production and post-production together, which had been separated up to this time. Today most production companies also do their own post.

The third advancement this brought us was the Non-linear editing (NLE or NLVE if you add the word video) or non-destructive editing. This also applied to digital audio workstations (DAW) for audio post. There had already been NLEs, but systems like Lucasfilm's EditDroid, the most popular in 1980, was very cumbersome and needed a lot of Laserdiscs to support editing.

There had even been NLEs as far back as the early 1970s, with system like the CMX, but the game changer came from Herb Dow, A.C.E who used the Ediflex that could use a bank of multiple Sony, JVC, and Panasonic Video Cassette Recorders (VCRs) as sources. After Herb Dow, A.C.E edited "Still the Beaver" in 1985, the system was adopted in about 75% of network programs and started to penetrate the film community, mostly for shorts and documentaries. Now you could go from filming to post and to air all on one medium.

In January 1984, Eastman Kodak announced the Video8 camera technology and in 1985, Sony introduced the HandyCam. This was an important step as it gave us the "Pro"-sumer camera, bringing a professional level of production to almost anyone. This allowed a lot more people to become filmmakers who just wanted to produce their own small productions and brought more interest in growing the independent community.

By late 1989 the first computer-based system, the EMC2, was released. One year later, Avid® showed off their system based on the Apple Macintosh. Even though it was only 15 Frames Per Second (FPS) and the Lucasfilm's EditDroid was much higher quality, the die was cast and the market embraced these technological advances.

This also gave us the codec (Coder/Decoder file) that let us deal with video formats in new ways, and almost all video and files use some kind of codec today. This is not to be confused with AVI that is sometimes mistakenly described as a codec: AVI is actually a container format. There are many container formats, like AVI, ASF, QuickTime, RealMedia, and Matroska. Within those containers there are many actual codecs like Divx, H.264, Sorenson and dozens more. These containers are packages that can contain multiple codecs, one each for audio, video and other playout sources.

The next advancement came after Disney engineers developed a long form solution that allowed the Macintosh to go beyond the 50GB storage limitation of that time. In April that year, Avid® and others introduced their new systems that could take advantage of the expansion of memory, and within two years the Avid® Media Composer had displaced most 35 mm film editing systems in most of the motion picture studios and TV stations worldwide. This made Avid® the undisputed authority in "off-line" non-linear editing systems. Even today the Avid® products are what people first think of when they talk about professional NLEs.

Apple set another bar by acquiring the software developed by macromedia engineers for the Media 100 hardware after Macromedia decided to get out of the video editing business. The Macromedia product evolved to be named "Final Cut" and Apple eventually released it as Final Cut Pro. This full featured, professional grade software was priced at only $1,499.00. This software was mostly revolutionary for its price, as it, like the Video8 and HandyCam, allowed small businesses or high schools to create their own projects and finish them. Until then, the Avid® tools started at $28,000.00 for just the software and basic breakout box. Final Cut Pro was also revolutionary as it was based on the QuickTime codec for media handling and it featured the ability to ingest video and audio via a computer firewire connection, so you did not need expensive cards or breakout boxes.

Companies like Adobe jumped on this workflow within a few years. Avid® decided to stay with their proprietary expensive hardware system and suffered as Apple and Adobe took more of Avid®'s market share as well as all of the millions, and now billions of entry level filmmakers. The advantages of a firewire workflow caused many camera manufacturers to produce new cameras and decks with firewire, grinding the barrier to entry even lower.

The next big change is when the Red One camera started shipping in August of 2007. RED lowered the cost of a high-end, digital cinema production camera, from that of the Panavision rig, at about $120,000 fully loaded, down to under $50,000. This also was the writing on the wall for film as the dominant production workflow. Thanks to RED, the professional independent films community started to really grow, now that the cost of equipment was somewhat affordable.

All the elements had now come together: we had professional NLEs, cameras that can shoot in high resolution and almost anyone could afford it. This triggered the next major innovation: large screen productions, which are still evolving and the only remaining use for 35mm.

In 2003, Sony introduced the tapeless video format media, the XDCAM, and the Professional Format Disc (PFD) to the world. Panasonic followed in 2004 with its P2 format, which used solid-state memory cards as a recording medium for DVCPRO-HD video. Then in 2006 Panasonic and Sony collaborated on the AVCHD format as an inexpensive, tapeless, high-definition video format. Today AVCHD camcorders are produced by Sony, Panasonic, JVC, Canon and others.

The Types of Cameras and Their Uses

The earliest video cameras were mechanical flying-spot scanners which were in use in the 1920s and 1930s during the period of mechanical television. Improvements in tube technology led to the development of video camera tubes in the 1930s and the underlying technology changed to support television broadcasting. Early designed cameras were extremely large devices, typically constructed in two sections, and often weighing over 250 pounds, too large for an operator to hand carry. The camera section held the lens and tube pre-amplifiers and other necessary electronics and was connected by a large diameter multicore cable to the remainder of the camera electronics, mounted in a separate studio room or a remote broadcast truck. Standalone, the camera head was unable to generate a video picture signal. Once created, video signals were output to the studio for transmission or recording. By the 1950s, solid-state electronics miniaturization had progressed to the point where some monochrome cameras could operate standalone and handheld. However, as a result of decades of applications, studio configuration remained static, with cameras connected via a large cable bundle to move the signals back to the camera control unit (CCU) in the studio or truck. The CCU was used to align and operate the camera's functions, including exposure, system timing, video and black levels.

By the 1950s in the U.S. and the 1960s in Europe, the first color cameras were introduced, and the system complexity increased as there were three or four pickup tubes, and their size and weight also increased. Hand-operated color cameras did not come into general use until the early 1970s with the first generation of cameras still split into a camera head unit, the body of the camera, containing the lens and pickup tubes, and held on the shoulder or a body brace in front of the operator, connected via a cable bundle to a backpack CCU. For field work, a separate Video Tape Recorder (VTR) was still required to record the camera's video output. "Camcorders" combine a camera and a VTR. Designed with extreme mobility in mind, these are widely used for television production, home movies, electronic news gathering (ENG), and similar applications. Since the transition to digital video cameras, most cameras have in-built recording media and essentially are camcorders. When these products were first introduced, the typical recorder was either a portable 1" reel to reel VTR, or a portable 3/4" U-matic VCR (Video Cassette Recorder). To operate the original Camcorder systems, the two camera units would be carried by the camera operator, while a tape operator would carry the portable recorder. By 1976, new product designs allowed camera operators to carry on their shoulders a one-piece camera containing all the electronics to output a broadcast quality composite video signal; however, a separate videotape recording unit was still required.

Due to the cost of producing and editing film, Electronic news-gathering (ENG) cameras replaced the 16mm film cameras for TV news production in the mid-1970s. Portable video tape production also enabled quicker response time for the completion of timely stories when compared to the need to chemically process film before review or edit.

Advances in solid-state technology lead to Charge-Coupled Device (CCD) imagers which were introduced in the mid-1980s. The first CCD cameras could not compete with the quality of color or resolution found in tube cameras of the same period, but the benefits of CCD technology, including smaller, lightweight cameras, a more stable image not prone to image burn-in or lag, and easy set-up calibrations meant development on CCD imagers quickly expanded its use in the industry. Once the quality of the images reached the level of the tube sensor, CCD cameras began displacing tube-based systems, which were almost completely retired by the advent of the 1990s. During the 1990s, cameras with the recorder permanently mated to the camera head became the norm for ENG. At the same time, in studio camera design, the camera electronics shrank as CCD imagers replaced the pickup tubes. The thick multicore cables connecting the camera head to the CCU were replaced by TRIAX connections, a slender video cable that carried multiple video signals, intercom audio, and control circuits, and could be run for a long distance, up to a mile if necessary. While the camera size reduced, and electronics no longer required over-sized housings, the typical "box" camera shape remained, as it must continue to hold large studio lenses, teleprompters, an electronic viewfinder (EVF), and other gear needed for studio and sports production. Electronic Field Production cameras are sometimes mounted in studio configurations inside a cage to support additional studio accessories.

In the late 1990s as High Definition Television broadcasting commenced, HDTV cameras suitable for news and general-purpose work were introduced. Delivering higher image quality, their operation was identical to their standard definition predecessors. At the turn of the century, new methods of recording for cameras were announced, including interchangeable hard-drives and systems based on flash memory which ultimately supplanted other forms of recording media.

Professional grade video cameras are designed for different purposes. Modern video cameras, like those used for television production, may be studio-based or designed for electronic field production (EFP). These systems generally offer detailed manual control of all parameters for the camera operator, sometimes to the exclusion of automated operation and usually use three sensors to separately record the red, green, and blue color images.

"Camcorders" combine a camera and a Video Cassette Recorder (VCR), a hard drive, or other recording device in a single package. Some action cameras have 360° recording capabilities and there are specialty systems for super speed capture and systems optimized for live event playback.

Closed-circuit television (CCTV) employs pan-tilt-zoom cameras (PTZ) for security, surveillance, and/or monitor requirements. Designed to be small, circumspect, and to support unattended operations when used in industrial or scientific settings, they are devised to support environmental factors typically inaccessible or uncomfortable for humans and are hardened for hostile environments (e.g., high radiation or heat, toxic chemical exposure, etc.). Webcams are video cameras which stream a live video feed to a computer. Camera phones have video cameras incorporated into mobile devises and recent quality enhancements of these platforms have led to some independent film productions relying on the mobile phones for creative effects. "Lipstick cameras" are named because the lens and sensor block combined are similar in size and appearance to a lipstick container: These miniature cameras are either hard mounted in a small location, like a race car, or on the end of a boom pole. The sensor block and lens are separated from the rest of the camera electronics by a long thin multi-conductor cable. The camera settings are manipulated from this box, while the lens settings are normally set when the camera is mounted in place.

The modern professional video camera, still referred to as a television camera even though the use has spread beyond broadcasting, is a feature full device for creating electronic moving images. In 2000s, major manufacturers such as Sony and Philips introduced completely digital professional video cameras. These cameras used CCD sensors and recorded video digitally on flash memory storage. These were followed by digital High Definition cameras (HDTV). As digital technology improved, supporting the transition to digital television transmissions, digital professional video cameras became dominant in television studios, Electronic News Gathering and EFP. With the advent of digital video capture in the 2000s, the distinction between professional video cameras and movie cameras disappeared as the internal technology and mechanism for capture became the same for both applications. For our purposes, mid-range cameras dedicated to television and professional media collection are termed professional video cameras.

Most professional cameras utilize an optical prism block directly behind the lens. This prism block (a trichroic assembly comprising two dichroic prisms) separates the image into the three primary colors, red, green, and blue, directing each color into a separate charge-coupled device (CCD) or Active pixel sensor (CMOS image sensor) mounted to the face of each prism. Some high-end consumer cameras also do this, producing a higher-resolution image, with better color fidelity than is normally possible with just a single video pickup. In both single sensor and triple sensor designs the weak signal created by the sensors is amplified before being encoded into analog signals for use by the viewfinder and also encoded into digital signals for transmission and recording. The analog outputs were normally in the form of either a composite video signal, which combined the color and luminance information to a single output, or an R-Y B-Y Y component video output through three separate connectors.

For our purposes, it is helpful to understand the differences between camera applications in their typical settings. Most television studio cameras stand on the floor, usually with pneumatic or hydraulic mechanisms called pedestals to adjust the height and are mounted on wheels. The CCU is connected via a TRIAX, fiber optic or the multicore cable, although these are very rarely used in today's studios. The CCU along with genlock and other equipment is installed in the production control room (PCR) often known as the "Gallery" of the television studio. When used outside a formal television studio in outside broadcasting (OB), they are often on tripods that may or may not have wheels. Studio cameras are light and small enough to be taken off the pedestal and the lens changed to a smaller size to be used on an operator's shoulder, or mounted on a dolly or a crane, making the cameras much more versatile than previous generations of studio cameras. These cameras are outfitted with a "tally light," a small signal-lamp that indicates, for the benefit of those being filmed as well as the camera operator, that the camera is "live" and its signal is being used for the "main program" when the light is lit.

Electronic News Gathering (ENG) video cameras were originally designed based on input and direction from news and field production camera operators. ENG cameras are larger and heavier than consumer hand-held models to dampen small movements and are typically supported by a shoulder support on the camera operator's shoulder, freeing a hand to operate the zoom lens control. These cameras can be mounted on tripods with fluid heads with a quick release plate and have interchangeable lenses. The lens is focused manually without intermediate servo controls. There are usually options to use a behind-the-lens filter wheel for selecting neutral density light filters. Accessible controls are implemented with hard, physical switches, usually in the same camera location, controlling Gain Select, White/Black balance, color bar select, and record start controls – these are not selected via software or menu selection to better support in field, manual operation. All settings, like white balance, focus, and iris can be manually adjusted, and automatic controls can be completely disabled to provide better creative management of the image. Professional grade BNC style connectors are supplied for video output and the "genlock" synchronization feed input, and a minimum of two professional XLR style input connectors are supplied for audio. Often there is a direct input for a portable wireless microphone, and audio is fully adjustable via easily accessed knobs. In addition to an electronic view finder, the video feed can be output to an external CRT viewfinder and as a professional tool it is equipped with a time code control section, allowing time presets. Multiple-camera setups can be time code-synchronized or "jam-synced" or forced to synchronize to a master clock. Usually these professional tools have "bars and tone" available in-camera, allowing the operator to insert the Society of Motion Picture and Television Engineers (SMPTE) color bars, a reference signal that simplifies calibration of monitors and levels setting when duplicating and transmitting the picture.

Electronic field production (EFP) cameras are similar to studio systems as they are used in multiple-camera switched configurations, but their main application is a deployment outside a controlled, studio environment, for concerts, sports, and live news coverage of special events. Versatility is the main advantage of these cameras, as they are designed to be flexible, carried on the shoulder or mounted on camera pedestals and cranes, and can support the large, very long focal length zoom lenses made for studio camera mounting. As opposed to camcorder ENG cameras, these cameras have no recording ability, and transmit their signals back to the broadcast truck through a fiber optic, TRIAX, or via a Radio Frequency link. Remote cameras are a separate type, typically very small camera heads designed to be operated by remote control. Despite a diminutive size, they are capable of performance comparable to the larger ENG and EFP cameras. "Block" cameras are named because the camera head is a small block, usually smaller than the lens. Sometimes completely self-contained, other block cameras only contain the sensor block and pre-amps requiring connection to a separate camera control unit in order to operate. All the functions of the camera can be controlled from a distance, including controlling the lens focus and zoom. Typically, these cameras are pan and tilt mounted, and can be placed in a stationary position, such as atop a pole or tower, in a corner of a broadcast booth, or behind a basketball hoop. Block cameras can also be placed on robotic dollies, at the end of camera booms and cranes, or "flown" in a cable supported harness as seen in sports productions where a camera has a bird's eye view of the field of play.

The Basics of Digital Cameras

The initial digital camera concept began with Eugene F. Lally of the Jet Propulsion Laboratory, who was applying mosaic photosensor technology to capture digital images. In 1961, his idea was to take pictures of the planets and stars while travelling through space to provide information to locate astronauts' position. Later in 1972, a Texas Instruments employee Willis Adcock had an idea for a filmless camera (U.S. patent 4,057,830). In both of these cases, the concept was good, but our technology was not ready to support the innovation. By 1975 a commercial all-digital camera called the Cromemco Cyclops was introduced. Originally a Popular Electronics hobbyist construction project, the design was published in the February 1975 issue of the magazine, and it used a 32×32 Metal Oxide Semiconductor sensor to capture images. Later that same year, 1975, an engineer at Eastman Kodak named Steven Sasson invented and built the first self-contained electronic camera that used a charge-coupled device (CCD) image sensor. Sasson's design was employed for military and scientific applications and by 1976 medical and news applications followed. As the professional cameras began to adopt digital technology, these ideas and technical innovations were applied to professional designs.

The two major types of digital image sensor are CCD and CMOS. A CCD sensor has one amplifier for all the pixels, while each pixel in a CMOS active-pixel sensor has its own amplifier. Because of this design, compared to CCDs, CMOS sensors use less power. While there are many arguments about the viability of the two methods, and which provides the best images, overall final image quality is more dependent on the image processing capability of the camera than on sensor type.

Resolution is important, whether the system is Standard Definition (SD), High Definition (HD), or Ultra High Definition (UHD). The resolution of a digital camera is often limited by the image sensor that turns light into discrete signals. The brighter the image at a given point on the sensor, the larger the value that is read for that pixel. Depending on the physical structure of the sensor, a color filter array may be used, which requires "de-mosaic-ing" to recreate a full-color image. The number of pixels in the sensor determines the camera's "pixel count." In a typical sensor, the pixel count is the product of the number of rows and the number of columns. For example, a 1,000 by 1,000-pixel sensor would have 1,000,000 pixels, or 1 megapixel.

Image sharpness is another measurement of camera performance. As you would expect, the final quality of any captured image depends on all optical transformations in the chain of producing the image. The weakest link in an optical chain determines the final image quality. In case of a digital camera, a simplistic way of expressing it is that the lens determines the maximum sharpness of the image while the image sensor determines the maximum resolution.

Let's take a deeper look into the heart of modern cameras, to better understand the methods employed to capture images and recognizing that each of these methods is generating more data. When digital innovations were first introduced, three different image capture methods were developed, and the hardware sensors and color filters adapted to support each method. The three types are single-shot, multi-shot, and scanning. Single-shot capture cameras use either one sensor chip with a "Bayer" filter, or three separate image sensors for the primary additive colors red, green, and blue that are exposed to the same image via a beam splitter. These were called three-CCD cameras. A "Bayer" light filter is an arrangement of color filters in a mosaic in the capture pixel array. The Bayer pattern is a repeating 2×2 mosaic pattern of filters, with green at opposite corners and red and blue in the other two positions. This means that green takes twice the proportion of filters and this was consciously designed to mimic the properties of our human eyes, which determine brightness from green and are more sensitive to brightness than to hue or saturation. Some cameras have used a 4-color filter pattern, typically adding two different hues of green to provide a potentially more accurate color but requiring complications in the interpolation process.

Multi-shot cameras were designed to expose the digital sensor to the image in a sequence of openings of the lens aperture. As there are a number of manufacturers that developed hardware using this method, the multi-shot technique has been applied in a number of different ways. Originally, the systems adapted a single image sensor to capture the same image as three filters were sequentially passed in front of the sensor to obtain additive color information. Later, a multiple shot innovation called micro-scanning was developed and this method employs one sensor chip with a color filter, typically a Bayer filter, and the hardware system physically manipulates the sensor in the focus plane of the lens to build a better resolved image than the native resolution of the chip supports. Latter developments combined both methods without a Bayer filter on the chip.

Scanning, as its name suggests, captures images by moving the sensor chip across the focal plane, in an operation like a document scanner. These cameras can offer very high-resolution images. Camera systems employ "linear" or "tri-linear" sensors in a single line of photosensors, or three lines for the three colors and the scanning operation is accomplished by physically moving the sensor or by rotating the whole camera. These digital rotating line cameras offer images of very high total resolution and the best color fidelity.

Single-shot systems with Bayer filters are typically consumer models and require an optical anti-aliasing filter to reduce aliasing due to the reduced sample rate of the different primary color images. Often the solution is a de-mosaic software algorithm to interpolate the color information and supply a full range of RGB color data. Professional cameras that use a beam-splitter single-shot 3 chip approach, three-filter multi-shot approach, color co-site sampling, or specialized sensors do not use anti-aliasing filters, nor de-mosaic software. Software in a raw converter program, such as *Adobe Camera Raw*, interprets the color data from the sensor to obtain a full color image, because the RGB color model requires three intensity values for each pixel: one measurement for red, green, and blue. Other color models also require three or more values per pixel. A single sensor element cannot simultaneously record these three intensities, and so a color filter array (CFA) must be used to selectively filter a particular color for each pixel. Using software, any color intensity values not captured for a particular pixel can be interpolated from the values of adjacent pixels which represent the color being calculated.

Since 2008, manufacturers have offered consumer DSLR cameras with a "movie" mode capable of recording high definition motion video. A DSLR with this feature is often known as an HDSLR or DSLR video shooter. Early HDSLRs capture video using a nonstandard video resolution or frame rate. The first DSLR introduced with an HD movie mode, the Nikon D90, captures video at 720p24 (1280×720 resolution at 24 frame/s). HDSLRs use the full camera imager area to capture HD video, though not all pixels which can introduce video artifacts. Compared with the higher resolution image sensors found in typical camcorders, HDSLR's much larger sensors yield distinctly different image characteristics. HDSLRs can achieve much shallower depth of field and superior low-light performance. Still, because of the low ratio of active pixels to total pixels, these consumer models are more susceptible to aliasing artifacts in scenes with particular textures, and the internal rolling shutter tends to introduce more artifacts. Because of the DSLR's optical construction, HDSLRs usually lack important video functions found on standard dedicated camcorders, such as autofocus while shooting, powered zoom, and an electronic viewfinder/preview screen. Additional limitations caused by their handling have prevented the HDSLR cameras from taking their place in the industry as simple point-and-shoot camcorders, requiring special planning and skills to gather professional quality images.

Over the past ten years, video functionality in these consumer models has continued to improve including higher video resolution and video bitrate, improved automatic control (autofocus) and manual exposure control, and support for formats compatible with high-definition television broadcast, Blu-ray disc mastering or Digital Cinema Initiatives (DCI). Models now offer broadcast compliant 1080p24 video. These developments have sparked a digital filmmaking revolution, and the "Shot On DSLR" badge of honor is becoming a mainstay for documentary and independent producers. An increased number of films, documentaries, television shows, and other productions are utilizing the quickly improving features.

Affordability and convenient size compared with professional movie cameras is driving rapid adoption.

The Implications of Camera Imaging Sensor Choice on System Requirements

Let's quickly review what we know about image sensors: An imaging sensor is a detection device that measures light and conveys the information that constitutes a total image. The sensors work by converting the variable attenuation of light waves as they pass through or reflect off objects into electrical signals, small bursts of current that convey the data. Early analog sensors for visible light were video camera tubes. Today the sensors are digital, either semiconductor "charge-coupled devices" (CCD), or active pixel sensors in "complementary metal–oxide–semiconductor" (CMOS) or "N-type metal-oxide-semiconductor" (NMOS, also called Live MOS) technologies. While CCD sensors are used for high-end broadcast quality video cameras, consumer products with internal cameras use lower cost and physically smaller CMOS sensors which offer lower power consumption in battery powered devices.

The fundamental operation of the CCD sensor is analog as each cell of a CCD pixel image sensor is an analog capture device. When light strikes the chip, it is retained as a small electrical charge in each sensor. Small output amplifiers are connected to a line of pixel sensors. The charges in the line of pixels nearest to the output amplifiers are amplified and output, then each line of pixels shifts its charges one line closer to the amplifier, filling the empty line closest to the amplifiers. This process is then repeated until all the lines of pixels have had their charge amplified and output. Alternatively, the CMOS image sensor has an amplifier for each pixel compared to the few amplifiers of a CCD. This results in less area for the capture of photons than a CCD, but this issue is typically resolved by using micro-lenses in front of each photodiode, which redirect light into the photodiode that would have otherwise bounced off the amplifier and be undetected. Some CMOS imaging sensors also use a method of back-side illumination to increase the number of photons that hit the photodiode. CMOS sensors can potentially be implemented with fewer components, use less power, and/or provide faster readout than CCD sensors, and most important for consumer products, they are also less vulnerable to static electricity discharges.

Many parameters can be used to evaluate the performance of an image sensor, including dynamic range, signal-to-noise ratio, and low-light sensitivity. For sensors of comparable types, the signal-to-noise ratio and dynamic range improve as the size increases.

Optical resolution describes the ability of an imaging system to resolve detail in the object that is being imaged. An imaging system may have many individual components including the lens the recording and the display components. Each of these contributes to the optical resolution of the system, as will the environment in which the imaging is done.

The optical transfer function (OTF) of an optical system such as a camera, microscope, human eye, or projector specifies how different spatial frequencies are handled by the system. The OTF is used by engineers to describe how the optical system projects light from the object or scene to the next item in the optical transmission chain, whether it is onto a photographic film, a detector array, a retina or screen. Some optical and electrical engineers prefer to use the modulation transfer function (MTF), an electrical measurement that neglects phase effects, but in most situations is the measured equivalent of the OTF. Either function specifies the response to a periodic sine-wave pattern passing through the lens system, as a mathematical function of its spatial frequency (period), and its orientation (phase). To get technical, the OTF is formally defined as the Fourier transform of the point spread function (PSF), which is the impulse response of the optics, the image of a point source. As a Fourier transform, the OTF is complex-valued; but it will be real-valued in the common case of a PSF that is symmetric about its center. The MTF is formally defined as the magnitude (absolute value) of the complex OTF. These spatial measurements allow us to quantify the sharpness of a diffraction-limited focused imaging system, or the blurriness of an out-of-focus imaging system, and they take into account important factors such as resolution.

Color separation is a function of the color image sensors employed in the device and is described by the three types of color-separation mechanism:

- Bayer filter sensor, using a color filter array that passes red, green, and blue light to selected pixel sensors. Each individual sensor element is made sensitive to green, red, and blue by means of a colored chemical dye gel placed over each individual element. Inexpensive to manufacture, this technique lacks the color purity of dichroic filters. Because the color gel segment must be separated from the others by a "freme," a separation like the caning in stained glass windows, less of the areal density of a Bayer filter sensor is available to capture light, making the Bayer filter sensor less sensitive than similar sized color sensors. In today's designs, most common Bayer filters use two green pixels, and one each for red and blue, which results in less resolution for red and blue colors, but this is acceptable because it correlates to the human optical system's reduced sensitivity at the edges of our visual spectrum. Missing color samples are interpolated using a de-mosaic algorithm or ignored altogether by a compression scheme. To improve color information, techniques like color co-site sampling use a hardware mechanism to shift the color sensor in pixel steps.

Foveon X3 sensor, using an array of layered pixel sensors, separates light through the wavelength-dependent absorption properties of silicon. This system allows every pixel location to sense all three colors and is similar to color photography film operations.

- 3-CCD, using three discrete digital image sensors, with color separation done by a dichroic optical prism. The dichroic prism elements provide sharper separation, improving overall color quality at full resolution. 3-CCD sensors produce better low-light performance and produce a full 4:4:4 signal, which is preferred in broadcasting, video editing, and chroma key visual effects. Every pixel can be measured with three color values, a Y value, a Cb value, and a Cr value. The maximum measurement for each of these is "4," so broadcasters and videographers use the shorthand of "4:4:4" to describe the maximum measurement for each parameter and this refers to the best possible color for an image.

Lens resolution is the ability of a lens to resolve detail and is determined by the quality of the lens limited by diffraction. Light coming from a point in the object diffracts through the lens aperture in such a way that it forms a pattern of diffraction in the image. This pattern will have a center spot and surrounding bright rings, separated by dark nulls, and this pattern is known as an Airy pattern, and the central bright spot as an Airy disk. Two adjacent points in the object give rise to two diffraction patterns. If the angular separation of the two points is significantly less than the Airy disk angular radius, then the two points cannot be resolved in the image, but if their angular separation is much greater than this, distinct images of the two points are formed and they can therefore be resolved. Only the very highest quality lenses have diffraction limited resolution, and normally the quality of the lens limits its ability to resolve detail. This ability is expressed in the aforementioned Optical Transfer Function which describes the spatial (angular) variation of the light signal as a function of spatial (angular) frequency. When the image is projected onto a flat plane, such as photographic film or a solid-state detector, spatial frequency is the preferred domain, but when the image is referred to the lens alone, angular frequency is preferred. OTF may be broken down into the magnitude and phase components in its mathematical measurement formulas and can be explained as spatial frequency in the x- and y-plane, respectively. OTF accounts for aberrations. The magnitude is known as the Modulation Transfer Function (MTF) and the phase portion is known as the Phase Transfer Function (PTF). In imaging systems, the phase component is typically not captured by the sensor making the important measure of the MTF.

Film, solid-state devices like CCD and CMOS detectors, and tube detectors (vidicon, plumbicon, etc.), as optical sensors detect spatial differences in electromagnetic energy. The ability of such a detector to resolve those differences depends mostly on the size of the detecting elements. Spatial resolution is typically expressed in line pairs per millimeter (lppmm), lines of resolution, mostly for analog video, contrast vs. cycles/mm, or MTF (the modulus of OTF). The MTF may be mathematically found by taking the two-dimensional Fourier transform of the spatial sampling function. Smaller pixels result in wider MTF curves and better detection of higher frequency energy. This is analogous to taking the Fourier transform of a signal sampling function: in that case, the dominant factor is the sampling period, which is analogous to the size of the picture element (pixel). Other factors include pixel noise, pixel crosstalk, substrate penetration, and fill factor.

A common problem among non-technicians is the use of the "number of pixels" on the detector to describe the resolution. If all sensors were the same size, this would be acceptable. But they are not, so the use of the number of pixels is misleading. For example, all else being equal, a 2-megapixel camera of 20-micrometer-square pixels will have worse resolution than a 1-megapixel camera with 8-micrometer pixels.

For a reliable resolution measurement, film manufacturers typically publish a plot of Response (%) vs. Spatial Frequency (cycles per millimeter). This measurement is derived experimentally. Solid-state sensor and camera manufacturers normally publish specifications from which the user may derive a theoretical MTF according to a recommended procedure. A few may also publish MTF curves, while others will publish the response (%) at the Nyquist frequency, or publish the frequency at which the response is 50%.

To find a theoretical MTF curve for a sensor, it is necessary to know three characteristics of the sensor: the active sensing area, the area encompassing the sensing area, the interconnection and support structures ("real estate"), and the total pixel count. The total pixel count is usually supplied, but as we have discussed it can be misleading. Sometimes the overall sensor dimensions are given, from which the real estate area can be calculated. Whether the real estate area is given or derived, if the active pixel area is not given, it may be derived from the real estate area and the fill factor, where fill factor is the ratio of the active area to the dedicated real estate area.

Time also impacts our sensor resolution. An imaging system running at 24 frames per second is essentially a discrete sampling system that samples a two-dimensional area. These systems also suffer from sampling limitations. For example, all sensors have a characteristic response time. Film is limited at both the short and the long resolution extremes, typically understood to be anything longer than 1 second and shorter than 1/10,000 second.

Additionally, film requires a mechanism to advance it through exposure, or a moving optical system to expose it. These limit the speed at which successive frames may be captured.

Digital systems have a different limitation. CCD is speed-limited by the rate at which the charge can be moved from a single pixel site to the next. CMOS has the advantage of having individually addressable pixel cells, and this has led to its advantage in the high-speed photography industry. Tube systems like vidicons, plumbicons, and image intensifiers have specific applications. The speed at which they can be sampled depends upon the decay rate of the phosphor used in the tube. For example, the P46 phosphor has a decay time of less than 2 microseconds, while the P43 decay time is on the order of 2–3 milliseconds. A tube image sensor with P43 phosphor is unusable at frame rates above 1,000 frames per second. Physical temperature impacts detectors as well. If objects within a scene are in motion relative to the imaging system, the resulting motion blur will result in lower spatial resolution. Short integration times will minimize the blur, but integration times are limited by sensor sensitivity. Important in our world of digital video distribution, motion between frames in media production has an impact on digital movie compression schemes (e.g., MPEG-1, MPEG-2). There are sampling schemes that require real or apparent motion inside the camera (scanning mirrors, rolling shutters) that may result in incorrect rendering of image motion, so sensor sensitivity and other time-related factors have a direct impact on spatial resolution.

In HDTV and VGA systems, the spatial resolution is fixed independently of the analog bandwidth because each pixel is digitized, transmitted, and stored as a discrete value. Cameras, recorders, and displays are selected so that the resolution is identical from digital camera to display. Analog systems are different, where the resolution of the camera, recorder, cabling, amplifiers, transmitters, receivers, and display may be completely independent and overall system resolution is controlled by the bandwidth of the lowest performing component in the chain. There are two methods by which to determine system resolution. The first is to perform a series of two-dimensional calculations, measuring first the image and the lens, then the result of that procedure with the image sensor, and continue through all components of the system. This is a complicated and time-consuming computation and must be performed anew for each object to be imaged. The other method is to transform each of the components of the system into the spatial frequency domain, and then to multiply the two-dimensional results. In this way a system response may be determined without reference to an image. The mathematical calculation will employ the Fourier transform. Although this method is considerably more difficult to comprehend conceptually, it becomes easier to use, especially when different design requirements or imaged objects are to be tested.

A variety of optical resolution measurement systems are available. Typical test charts for Contrast Transfer Function (CTF) consist of repeated bar patterns. The limiting resolution is measured by determining the smallest group of bars, both vertically and horizontally, for which the correct number of bars can be seen. By calculating the contrast between the black and white areas at several different frequencies, however, points of the CTF can be determined with a mathematical contrast equation. In broadcast plants, some modulation may be seen above the limiting resolution; these may be aliased and phase-reversed to correct. When using other methods, including the interferogram, sinusoid, and the edge in the ISO 12233 target, it is possible to compute an entire MTF curve. The response to the edge is similar to a step response, and the Fourier Transform of the first difference of the step response yields the MTF. Other measurement systems include an interferogram created between two coherent light sources that can be used for at least two resolution-related purposes, the first to determine the quality of a lens system, and the second to project a pattern onto a sensor

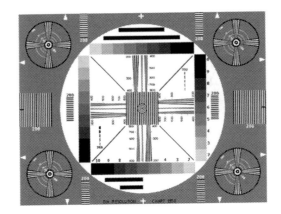

FIGURE 3.1 EIA 1956 Video Resolution Target

to measure resolution. The EIA 1956 resolution target, shown in Figure 3.1, and the similar IEEE 208–1995 resolution target, were specifically designed to be used with television systems.

The gradually expanding lines near the center are marked with periodic indications of the corresponding spatial frequency. The limiting resolution is found directly through inspection. The most important measure is the limiting horizontal resolution since the vertical resolution is typically determined by the applicable video standard. The ISO 12233 target was developed for digital camera applications, since modern digital camera spatial resolution may exceed the limitations of the older targets. It includes several knife-edge targets for the purpose of computing MTF by Fourier transform. They are offset from the vertical by 5° so that the edges will be sampled in many different phases, which allow estimation of the spatial frequency response beyond the Nyquist frequency of the sampling.

Finally, a Multi-burst signal is an electronic waveform used to test transmission, recording, and display systems. The test pattern consists of several short periods of specific frequencies. The contrast of each may be measured by inspection and recorded, giving a plot of attenuation vs. frequency. The NTSC3.58 multi-burst pattern consists of 500 kHz, 1 MHz, 2 MHz, 3 MHz, 3.58 MHz, and 4.2 MHz blocks. 3.58 MHz is important because it is the chrominance frequency for NTSC video. PAL Multiburst packets are 0.5 MHz, 1.0 MHz, 2.0 MHz, 4.0 MHz, 4.8 MHz, 5.8 MHz as per CCIR 569.

THE VARIOUS FORMS OF COLOR CODING & COLOR SPACES

Color space is a specific organization of colors. When used in combination with the physical parameters of a specific device, it supports in both analog and digital representations identical representations of color. Color space may be defined by individual interpretation, for example the Pantone collection where particular colors are assigned to a set of physical color swatches and assigned names or numbers. Alternatively, color space can be structured mathematically, as defined by the NCS System, Adobe RGB or sRGB. Color models are abstract mathematical models describing the way colors can be represented as tuples of numbers (e.g., triples in RGB or quadruples in CMYK), yet a color model with no associated mapping function to an absolute color space becomes an arbitrary color system with no connection to any internationally recognized standard of interpretation. By employing a specific mapping function between a color model and a reference color space, media producers establish within the reference color space a fixed, absolute gamut, or "footprint," and for a given color model this defines a color space. For example, Adobe RGB and sRGB are two different absolute color spaces, both based on the RGB color model. When defining a color space, the most used international reference standards are the CIELAB or CIEXYZ color spaces, which were specifically designed to encompass all colors an average human can see.

How did we arrive at the idea of a color space? Starting in 1802, Thomas Young posed the existence of three types of photoreceptors in the eye, which we now call "cone cells," each of which was sensitive to a particular range of visible light. By 1850, Hermann von Helmholtz developed the Young–Helmholtz theory that the three types of cone photoreceptors could be classified as short-preferring (blue), middle-preferring (green), and long-preferring (red), according to their response to the wavelengths of light striking the retina. The relative strengths of the signals detected by the three types of cones are interpreted by the brain as a visible color. The color-space concept followed and was introduced by Hermann Grassmann, who developed it in two stages. First, he developed the idea of vector space, which allowed the algebraic representation of geometric concepts in dimensional space. With this conceptual background, in 1853 Grassmann published a theory of how colors mix. This theory and its three-color laws are still taught as Grassmann's laws. As noted by Grassmann

> … the light set has the structure of a cone in the infinite-dimensional linear space. As a result, a quotient set of the light cone inherits the conical structure, which allows color to be represented as a convex cone in the 3- D linear space, which is referred to as the color cone.

Colors can be created with color spaces based on the CMYK color model, using the subtractive primary colors of pigment (cyan (C), magenta (M), yellow (Y), and black (K)). To create a three-dimensional representation of a given color space, we can assign the amount of magenta color to the representation's X axis, the amount of cyan to its Y axis, and the amount of yellow to its Z axis. The resulting 3-D space provides a unique position for every possible color that can be created by combining those three pigments. Colors can be created on monitors with color spaces based on the RGB color model, using the additive primary colors (red, green, and blue). A three-dimensional representation would assign each of the three colors to the X, Y, and Z axes. Note that colors generated on given monitor are limited by the reproduction medium, such as the phosphor (in a CRT monitor) or filters and backlight (LCD monitor). Another way of creating colors on a monitor is with an HSL or HSV color space, based on hue, saturation, brightness (value/brightness). With such a space, the variables are assigned to cylindrical coordinates. Many color spaces can be represented as three-dimensional values in this manner, as show in Figure 3.2, but some have more or fewer dimensions, and some arbitrarily defined color spaces, such as Pantone cannot be represented in this way at all.

Additive color mixing: Three overlapping lightbulbs in a vacuum, adding together to create white, as shown in Figure 3.3.

Subtractive color mixing: Three splotches of paint on white paper, subtracting together to turn the paper black.

RGB uses additive color mixing, because it describes what kind of light needs to be emitted to produce a given color. RGB stores individual values for red, green, and blue. RGBA

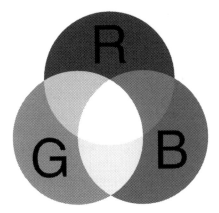

FIGURE 3.2 Additive Color Mixing

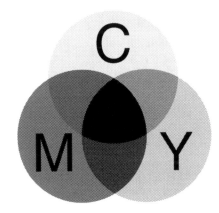

FIGURE 3.3 Subtractive Color Mixing

is RGB with an additional channel, alpha, to indicate transparency. Common color spaces based on the RGB model include sRGB, Adobe RGB, ProPhoto RGB, scRGB, and CIE RGB.

YIQ was formerly used in analog NTSC television broadcasts (National Television Standards Committee, used in North America, Japan, and elsewhere) for historical reasons. This system stores a "luma" value roughly analogous to, and sometimes incorrectly identified as, luminance, along with two chroma values as approximate representations of the relative amounts of blue and red in the color. It is similar to the YUV scheme used in most video capture systems and in PAL (Australia, Europe except France and Russia) television, except that the YIQ color space is rotated 33° with respect to the YUV color space and the color axes are swapped. The YDbDr scheme used by SECAM analog television in France and Russia is rotated in another way. YPbPr is a scaled version of YUV and it is most commonly seen in its digital form, YCbCr, used widely in video and image compression schemes such as MPEG and JPEG.

xvYCC is a new international digital video color space standard published by the IEC (IEC 61966-2-4). Based on the ITU BT.601 and BT.709 standards, it extends the gamut beyond the R/G/B primaries specified in those standards. HSV (hue, saturation, value), also known as HSB (hue, saturation, brightness) is often used by artists because it is often more natural to think about a color in terms of hue and saturation than in terms of additive or subtractive color components. HSV is a transformation of an RGB color space, and its components and colorimetry are relative to the RGB color space from which it was derived. HSL (hue, saturation, lightness/luminance), also known as HLS or HSI (hue, saturation, intensity), is quite similar to HSV, with "lightness" replacing "brightness." The difference is that the brightness of a pure color is equal to the brightness of white, while the lightness of a pure color is equal to the lightness of a medium gray.

WHAT DOES GAMMA AND LOG PROCESSING MEAN?

Gamma, the shortened and common reference for gamma correction, is the non-linear operation used to encode and decode luminance or tristimulus values in video or still image systems. In the simplest cases, gamma correction is defined by a mathematical power-law expression. A gamma value is sometimes called an encoding gamma, and the process of encoding with this compressive logarithmic expression is called gamma compression. In the reverse function, the gamma value is called a decoding gamma and the application of the expansive power-law logarithmic expression is called gamma expansion.

Gamma encoding of images is used to optimize the usage of data bits when encoding an image, or the bandwidth required to transport an image, by taking advantage of the non-linear way humans perceive color and light. With no actual relationship to the gamma function, the human perception of brightness in everyday situations, in other words not in darkness or blinding sunlight, follows an approximate logarithmic function with greater sensitivity to relative differences between darker tones than between lighter ones. We apply gamma correction because if images are not gamma-encoded, they allocate too much data or too much bandwidth to highlights that humans cannot differentiate, and too few bits or too little bandwidth to shadow values that humans typically do see and would require more bits/bandwidth to maintain an optimum visual experience.

Gamma encoding was originally developed to compensate for the input–output characteristic of cathode ray tube (CRT) displays. Light intensity varies nonlinearly with the electron-gun voltage in CRT displays. Altering the input signal by gamma compression can cancel this nonlinearity, such that the output picture has the intended luminance. The similarity of CRT physics to the inverse of gamma encoding needed for video transmission was a combination of coincidence and engineering, and simplified the early television set electronics. The advantage to modern systems is different. The gamma characteristics of the display device are less of a concern in the gamma encoding of images and video; we employ gamma encoding to maximize the visual quality of the signal, regardless of the gamma characteristics of the display device.

Until the recent advent of High Dynamic Range (HDR) televisions and monitors, most video screens were not capable of displaying the dynamic range of brightness that can be captured by typical digital cameras. Considerable artistic effort has been invested in choosing the reduced form for the original image to be presented. Contrast selection via the gamma correction is part of the artistic adjustments used to fine-tune the image for reproduction. Also, it is important to note that digital cameras record light using electronic sensors that respond linearly, not logarithmically like our eyes. The process of rendering linear raw data to conventional RGB data (e.g., for storage into JPEG image format), color space transformations and rendering transformations will be performed. In particular, almost all standard RGB color spaces and file formats use a gamma compression non-linear encoding of the image intensities of the primary colors and the intended reproduction is almost always nonlinearly related to actual measured scene intensities.

Binary data in still image files, such as JPEG photographic images, are explicitly encoded which means that they carry gamma-encoded values, not linear intensities, as are motion picture files compressed with the MPEG standard. The gamma encoding system can be tuned to manage both cases through color management, if a better match to the output device gamma is required. The sRGB color space standard used with most cameras does not use a simple power-law nonlinearity as described but has a decoding gamma value near 2.2 over much of its range. Below a compressed value of 0.04045 or a linear intensity of 0.00313, the curve is linear, in other words, the encoded value is directly proportional to intensity. Output to CRT-based television receivers and monitors does not usually require further gamma correction, since the standard video signals that are transmitted or stored in image files incorporate gamma compression that provides a pleasant image after the gamma expansion of the CRT. For television signals, the actual gamma values are defined by the video standards (ATSC or DVB-T) and are always fixed and published values.

WHAT RAW MEANS AND HOW TO DIGITALLY PROCESS IT

A camera RAW image file contains minimally processed data from the image sensor of either a digital camera or motion picture film scanner. RAW files are named because they are not yet processed and not ready to be printed or edited. Typically, raw images are processed by a RAW converter in a wide-gamut internal color space where precise adjustments can be made before conversion to a house preferred or "mezzanine" file format for storage or further manipulation. This usually encodes the image in a device-dependent color space. There are hundreds of RAW formats in use by different models of digital equipment and many are not compatible.

The purpose of RAW image formats is to save, with minimum loss, all data obtained from the sensor, as well as the metadata describing the conditions of the capture of the image. RAW image formats are intended to capture as closely as possible the complete characteristics of the scene, all pertinent physical information about the light intensity and color. Like photographic negatives, RAW digital images may have a wider dynamic range or color gamut than the eventual final image format as it maintains the most complete record of the captured image. Most RAW image file formats store information sensed according to the geometry of the sensor's individual photo-receptive elements, the pixels, rather than points in the expected final image: for example, camera sensors with hexagonal element displacement record information for each of their hexagonally displaced cells, which a decoding software will eventually transform into rectangular geometry during "digital developing." The process of converting a RAW image file into a viewable format is sometimes called "developing" a RAW image, harkening back to the film development process used to convert photographic film into viewable prints. Setting white balance, color grading and gamma all participate in the rendering process.

RAW files contain the information required to produce a viewable image from the camera's sensor data. The structure of RAW files often follows a common pattern:

- A short file header which typically contains an indicator of the byte-ordering of the file, a file identifier and an offset into the main file data.
- Camera sensor metadata, which is required to interpret the sensor image data, including the size of the sensor, the attributes of the CFA and its color profile.
- Image metadata which is required for inclusion in any CMS environment or database. These include the exposure settings, camera/scanner/lens model, date (and, optionally, place) of shoot/scan, authoring information and other. Some RAW files contain a standardized metadata section with data in Exif format.
- In the case of motion picture film scans, either the timecode, keycode or frame number in the file sequence which represents the frame sequence in a scanned reel. This item allows the file to be ordered in a frame sequence (without relying on its filename).
- The sensor image data: RAW files contain the full resolution data as read out from each of the camera's image sensor pixels.

If RAW format data is available, it can be used in high-dynamic-range imaging conversion, as a simpler alternative to the multi-exposure HDI approach of capturing three separate images, one underexposed, one correct and one overexposed, and "overlaying" one on top of the other.

To be viewed or edited, the output from a camera's image sensor has to be converted to a photographic rendering of the scene and stored in a standard format. This processing, whether done in-camera or later in a RAW-file converter, involves a number of operations, including but not limited to:

- decoding – image data of RAW files are typically encoded for compression purpose, but sometimes for security obfuscation purpose
- defective pixel removal – replacing data in known bad locations with interpolations from nearby locations
- white balancing – accounting for color temperature of the light that was used to take the photograph
- noise reduction – trading off detail for smoothness by removing small fluctuations
- color translation – converting from the camera native color space defined by the spectral sensitivities of the image sensor to an output color space (typically sRGB for JPEG)
- tone reproduction – the scene luminance captured by the camera sensors and stored in the RAW file (with a dynamic range of typically ten or more bits) needs to be rendered for pleasing effect and correct viewing on low-dynamic-range monitors or prints; the tone-reproduction rendering often includes separate tone mapping and gamma compression steps.
- compression – for example JPEG compression

Cameras and image processing software may also perform additional processing to improve image quality, for example:

- removal of systematic noise – bias frame subtraction and flat-field correction
- dark frame subtraction

- optical correction – lens distortion, chromatic aberration, and color fringing correction
- contrast manipulation
- dynamic range compression – lighten shadow regions without blowing out highlight regions

COMPRESSED FILE SYSTEMS

Video, once it is digitized, must be saved in a representative manner so that the content can be stored in an archive or transmitted as a file or a data stream. A video compression format is the coding that is applied to the content data to make files and there are thousands of types, from consumer to "pro"-sumer to professional methods. Video encoding to compressed formats can be "lossless," meaning that all the data is saved resulting in a very large file, or "lossy" where the data is compressed for smaller file size or lower transmission bandwidth, resulting in degradation of the original content data. Consumer media is usually compressed using "lossy" video codecs, resulting in significantly smaller files than lossless compression. While most video coding formats are designed either for lossy or lossless compression, some formats support both. Uncompressed video formats are a type of lossless video processing used in some circumstances such as when sending video to a display over an HDMI connection. Some high-end cameras can capture video directly in an uncompressed, lossless format. A few examples of video coding formats include MPEG-2 Part 2, MPEG-4 Part 2, JPEG2000, H.264 (MPEG-4 Part 10), DV-DVCpro, AVC-Intra, DNXHD, DPX, HEVC, Theora, RealVideo RV40, VP9, and AV1. A specific software or hardware implementation capable of video data compression and/or decompression to/from a specific video coding format is called a video "codec," and the operation of changing one format to another is called "transcoding."

It is important to know that there are hundreds of formats and variations in use in the media industry. While many codecs have been developed for specific software and hardware platforms, such as PRORES for the Apple Final Cut Pro Software or DNXHD for the AVID® Editing systems, most professional coding formats are documented in a detailed technical specification. Some specifications have been approved by standardization organizations and are considered a video coding standard. Because of the proliferation of codec formats, the use of the term 'standard' applies to both formal standards and de facto product promoted standards. Conceptually, there is a difference between a format "specification" and its codec implementations. Video coding formats should be detailed in specifications, and the software or hardware to encode and decode data to or from uncompressed video data are codec implementations of those specifications. This is one of the reasons why the industry struggles with hundreds of versions of the same format: for each specification, there can be many codecs implementing that specification in slightly different manners, often to promote a specific product feature or benefit.

In many cases, video content data encoded using a coding format is bundled with audio channels each encoded with an audio coding format, as well as subtitle or captions data, and "wrapped" inside a multimedia container format such as MXF (SMPTE Specification 386M, 383M, 381M & 377 are frequently used), Quicktime, AVI, MPEG4, FLV, or RealMedia. These file packages are a container holding the pieces of the total media asset and can be easily transmitted or stored. Wrapper multimedia container formats can contain any one of a number of different video coding formats, which complicates this situation for companies that share media files on a regular basis like broadcasters, production and post-production houses. When a user analyzes a new file, an MP4 wrapper container can contain compressed video made with the MPEG-2 Part 2 codec, or video compressed with the H.264 video coding format, or any of a number of other format options.

A video coding format specification does not dictate all algorithms used by a codec implementing the format. For example, a large part of how video compression typically works is by finding similarities between video frames and achieving compression by copying previously coded similar sub-images or "macroblocks" and adding small differences as they occur. In the real world it is almost impossible to find an optimal compression solution, but many manufacturers have developed hardware and software tools to manage the predictions of changes in the macroblocks as well as capture the differences. Since the video coding format standards do not dictate the algorithms that manage the encoding steps, systems can innovate in their support for compression across the video frames in the data stream. The application of the video often requires codec decisions that trade storage space versus time, in other words a live feed for broadcast quality may need a codec that operates fast but is very inefficient in file size and requires higher storage space, while a DVD codec needs to manage space on the storage medium and trades the speed of operation for higher compression and lower storage requirements.

The equipment in your plant will dictate the type of format your company requires. Choices are impacted by your editing software, your transmission chain, storage and Media library selections. One of the most widely used video coding formats is H.264. H.264 is a popular choice for encoding for Blu-ray Discs and it is widely used for "proxy" video that media asset management systems use to reference high definition files. It is also widely used by streaming Internet sources, like YouTube, Netflix, Vimeo, and the iTunes Store, web software such as the Adobe Flash Player and Microsoft Silverlight, and also various HDTV broadcasts over terrestrial (ATSC standards, ISDB-T, DVB-T, or DVB-T2), cable (DVB-C), and satellite (DVB-S2). Standards vying to be the next generation video coding format appear to be JPEG2000, the heavily patented HEVC (H.265) and AV1.

A subclass of video coding formats are the intra-frame formats, which apply compression to each picture in the video-stream in isolation, with little or no attempt to take advantage of correlations between successive pictures over time for a higher level of compression. An example is Motion JPEG, which is simply a sequence of individually JPEG-compressed images. These codecs tend to operate faster but build much larger files than a video coding format supporting interframe coding. Because interframe compression copies data from one frame to the next, should an originating frame be lost or damaged, the subsequently following frames cannot be properly reconstructed. Editing a video file compressed with interframe formats is difficult due to the fact that an original frame may not be at the selected edit point. Making edits in intraframe-compressed video is almost the same operation as editing uncompressed video. Another difference between intraframe and interframe compression is with intraframe systems, each frame uses a predictable and similar amount of data; however, in most interframe systems, particular frames – like the "I frames" in an MPEG-2 file – cannot copy data from previous frames, requiring larger amount of data than nearby frames. This is why you will hear some producers ask to set their MPEG codec for "all I-Frames," in fact emulating an intraframe codec to make smooth editing easer.

Video coding formats can define additional restrictions to be applied to encoded video, called profiles and levels. A "profile" restricts which encoding techniques are allowed and a "level" is a restriction on parameters such as maximum resolution and data rates. It is possible to have to supply a decoder which only supports decoding a subset of profiles and levels for a given video format. This is most often set to make the decoder program or hardware easier to use, faster in operation, or control the size of the encoded output file.

UNCOMPRESSED FILE SYSTEMS

"Uncompressed" video is digital video that has never been compressed or was generated by decompressing previously compressed media. It is most often found in video cameras, video monitors, some video recording devices and in video processors that manage operations like image resizing, deinterlacing, and text and graphics overlay. Uncompressed video can be transmitted over various types of baseband digital video interfaces like SDI. Some High Definition cameras feature the output of high-quality uncompressed video, while others compress the video using a lossy compression method, usually to achieve a lower price point. In a lossy compression process video information is removed, which leads to compression artifacts and reduces the quality of the resulting decompressed video. If your business is based on high-quality editing of digital video, it is best to work with video that has never been compressed or used lossless compression as this maintains the best possible quality. Compression can always be applied after editing chores are finished.

COMMON CAMERA MEDIA TYPES

As we learned from our history of the development of digital cameras, professional digital cameras began using back-mounted hard drives and we still find these capture hard drives still in use today, with some vendors adopting the SD card for firmware updates and small file transfers. Most professional camera manufacturers such as Sony, ARRI, Cannon and RED provide their own capture media as a part of their extended product line. These media types, despite standards are not interchangeable.

COMMON FILE FORMATS FOR CAMERAS

Common acquisition file formats in digital cameras include Uncompressed RAW, Apple ProRes, AVID® DNX, and MPEG. For camera work that is captured for immediate delivery or destined for delivery applications, there are a number of supported formats including for 4K and 2K images DPX and TIFF, and for High Definition and Standard Definition there is support for Apple Quicktime, JPEG, AVID® AAF, MXF, H.264, and MP4. The format is usually selected to support the editing and processing tools in the studio or facility. For example, if your production chain uses the AVID® editing systems or the Apple Final Cut Pro editing tools then the camera output format is chosen to match the in-house system.

All cameras provide metadata about the picture. Data may include aperture, exposure time, focal length, date and time taken, and location. Camera cards from different manufacturers have different file conventions, including folder structure and metadata storage, which often makes the media non-interoperable.

COMMON PRODUCTION FORMATS

There are primary production formats used in the industry today and they are better known by their manufacturer rather than the myriad of formats or product names: Avid®, Apple, and Adobe, the three "A's" of media and broadcast production.

AVID®—The **DNXHD** video coding format, and its many permutations over the years, is the AVID® specified format for editing and processing. The AVID® Media Composer has been a mainstay in Hollywood for over 20 years, and it includes a complete eco-system of network products (ISIS, etc.), storage products, production asset management (PAM) tools (Interplay, Media Central UX), audio tools like PRO TOOLS, newsroom automation systems (iNEWS) and graphics editing systems. AVID® offers dialogue search tools, a full featured RESTful API and support from a global distribution network.

Apple—**ProRes** is the video coding format introduced by Apple Computers to support high quality editing on their Final Cut Pro platform. In the early 1990s, the relatively inexpensive Final Cut Pro software and similarly priced Apple hardware platform to support it provided hundreds of operations a low cost, high quality video editing tool, lowering the financial barrier to entry for many small producers, broadcasters and post-production houses.

Adobe Systems—Touted to work with any camera uncompressed **RAW** files, the Premiere Pro video editing suite, coupled with the other tools in the Adobe Creative Cloud like After Effects for graphics, Audition for audio, and its internal Adobe Media Encoder. The media encoder is a versatile tool that transcodes from / to many different formats and variances in formats, and over the past 5 years has become a popular production software system, replacing many Final Cut Pro applications in the industry.

WHAT INGEST MEANS AND HOW TO ORGANIZE IT

Ingest is the process of capturing incoming media streams or files, analyzing them and bringing them into a system for management or further processing. Whether the media arrives from a camera card, an FTP site, or a content distribution accelerated network like IBM Aspera, Signiant or File Catalyst, or delivered over an ethernet network, the digitized assets must be recognized, analyzed, and published in the internal systems to enable further use.

Ingest processes are often described as a "workflow" as each company has a set of steps it uses to evaluate, catalog and integrate new digital media as it arrives. The ingest process typically employs tools such as digital video codecs to normalize the media to company specifications. The following list defines one example of an ingest process:

- Create metadata placeholder(s) – this is an optional step that many systems employ to prepare a pre-arrival metadata description of expected incoming digital media, typically with an identifier like a title, a house number or code, or a Universal Unique Identifier (UUID)
- Search/locate file(s) – files that have arrived can automatically trigger a "watch folder" to start the process, or the system can be commanded to seek the digital media files and start the process
- Pre-analysis of the High-Resolution file – what are its technical parameters? Does it meet house requirements? Is there metadata accompanying the media file(s) in a sidecar file or embedded within the "wrapper?" Is this an expected secured delivery or is there a virus scan to run on the data package?

- Map Media format – the ingest process must use the analysis to separate the components inside the wrapper and prepare its tools, such as codecs, to organize the work to be performed on the incoming file(s)
- Convert Media to house format – some companies and some systems require their media to be in a single format for use in their operations. This format is referred to as a "mezzanine" format and all incoming media must match it or be transcoded to the mezzanine format
- Media Proxy generation – if the media is to be used in a library management system, these systems typically create a low-resolution copy of the media to use as a reference to the high-resolution system for media annotations, edit points, and as an aid in quality control operations. These "proxies" substitute for the actual media for many internal functions and provide fast access for manual chores
- Create XML representation – define the incoming asset in a standardized way in a language that can be both human and machine readable, and can be used by asset, content, or production management software
- Catalog asset – place the XML representation in the system of record for the organization, whether it is asset, content, or a production management operation
- Miscellaneous Map Translator – any additional files, media "maps," or ancillary files found in the wrapper need to be translated into the organization's system for historical reference or operations if the system uses metadata-driven workflows
- Store High-Resolution instance – the new media asset and any associated components like audio files and caption files, must be stored in the most appropriate storage repository
- Create Index of High-Resolution asset – collect any technical information on the asset as well as any indicated points in the digital file, such as "Start of Media (SOM)," "End of Media (EOM)," etc.
- Generate storyboard – some systems need a series of still images to be generated for each asset to aid in user searchability and downstream processing of the media
- Set locations on Proxy – if the indexed Hi-res asset has points, set those indications on the proxy version of the original file
- Insert proxy audio – attach the audio channels to the video proxy
- Insert proxy subtitles or closed captions – attach the subtitle or caption files to the proxy
- Generate "Title" – in the system of record, there will be some internal name for the new asset, and this is the step that registers the new name, whether it is a title, a house number or some other form of tracking mechanism
- Report instance UUID – if the internal system of record is employing UUIDs (Universal Unique Identifiers), this is the step to register that tracking sequence
- Auto-tag with metadata and distribute incoming content – some sequences recognize media for particular purposes and can "tag" the new asset with a metadata marker to trigger a downstream workflow to deliver the new asset to a particular location, department, or user group
- Auto assign "new asset" tasks to internal staff – some systems move newly ingested assets into a production or quality review workflow and the ingest process can assign work based on the arrival of new media
- Escalation procedures – should the incoming digital file fail the process, the ingest workflow can trigger messages or alarms to notify the supplier or management to the issues of the failure

- Automated Quality Control (AQC) – many organizations use software tools to analyze the digital video and perform a quality review with a posted analysis report; some systems even annotate the proxies with the information from the report to aid in quick human review of the failure points
- Manual QC – typically this step is triggered by failure in an auto quality control review, but some companies insist on a human review of incoming media to check its viability
- Supplemental files/linking – for companies that are distribution Video on Demand (VOD) and Over the Top (OTT) service (HULU, Netflix, Verizon, etc.) versions, especially for companies that are distributing international language versions of their programming, the need to match and link supplemental files like audio and subtitle translations or an edited version to the original asset is key to the success of the business
- Unknown Media workflows – what to do with an asset that is unrecognizable? The ingest system needs an "escape valve" to pass the media to a human review process to correctly catalog and link the asset in the system of record
- Manager review steps – some organizations require management to review incoming media of particular types and these steps can be automatically organized by the ingest process
- Supplier analysis reports – if your organization is accepting media from syndicators or ad agencies, the ingest process may be required to save audit data on media sources for reporting or dashboard monitoring
- "Refused" media reports and supplier notifications – when media fails, especially if the failures are regular and caused by the same source, the ingest process is often required to notify the original supplier of the failures and provide management reports of the media failures.

This list is not an exhaustive list of steps for an ingest process. If the in-house system is supporting SMPTE Interoperable Master Format (IMF) for distribution, there may be additional steps including the reading of the IMF packing lists and applying it to the media map, noting any IMF mark-up points and indicating those on the proxy annotations, etc. Ingest workflows are often customized to the internal requirements of the organization as well as the needs of the software and hardware tools in use.

ASSET MANAGEMENT

Media Asset Managers or MAMs are our modern libraries for digital media files. They come in many different forms, many focused on particular roles in an organization and they are not to be confused with DAMs, Digital Asset Managers, the systems focused on document management which usually boast a completely different set of features and applications than their media counterparts. MAMs are designed specifically for media management and because modern metadata can drive workflows and automate operations, today's products are often deeply integrated to workflow orchestration systems and business process managers.

Asset management systems have become specialized, with systems managing specific workflows for archive and preservation operations, production support, media preparation for play-to-air for broadcast networks and television stations, syndication product distribution, news operations support, sports and live event production and "versioning" for international distribution, or to support Video on Demand (VOD) and Over-the-Top (OTT) services like Netflix, HULU, YouTube, etc. It is key to recognize that specialized MAMs do not necessarily address every operational requirement. Cost-effective systems support small operations while multi-site, geographically spanning enterprise systems can organize and manage global operations.

Core to any MAM is its organization of the media and components in the library and the systems searchability and ease of use. The library catalog for each asset is based on a set of metadata usually built upon a data model adapted to the needs of a particular application or a particular company's unique data collection requirements. MAMs feature media manipulation tools to playback and annotate proxies. Some orchestration systems use the metadata collected to drive workflows without human interaction. As an orchestration workflow engine, MAM systems must integrate into various third-party tools to complete the steps of the operational path. Recently there has been a move to add value by enhancing workflow operations with Machine Learning software tools, typically called "Artificial Intelligence" or AI. Clever AI applications are being introduced to lower the manual labor, especially in regard to reducing human effort and increasing efficiencies such as automatically applying metadata annotations to media files in the library.

The digital media ingest process is usually designed to be managed by the asset management system, and the automatic harvesting of metadata and technical data is key to the tool's success. When considering MAM systems, pre-planning and analysis of needs will go a long way in selecting the correct tool for a particular application as not all MAMs work the same way or provide the same results, nor do component systems all interconnect and function well together in a workflow orchestration system.

ON-SET GRADING

On-set grading is a technique used by cinematographers in which a certain "look" or visual style is applied to video or film material by the means of set lighting. In modern film and television production, since the images are captured in a digital format the method of applying a visual style is by applying color correction or color grading. These are artistic concerns that have implications throughout the production phases of a program. For example, incoming RAW camera footage can be color graded to a specific quality of light early in the capture process, but later in the production process when other processing is added to the image, it may be found to be too warm or too cool, in other words too red-yellow or too blue for the director's preference or artistic vision. This is especially apparent in modern film making when computer-generated imagery (CGI) character generation is added to productions. The cinematographers can go back to the original RAW footage and re-grade the digital file to adjust the color grading, so that the later processing maintains the image style that the director seeks.

DAILIES

What happened yesterday? This is the real question "dailies" answer. Dailies are the unedited camera footage collected during the making of a production, whether it be a television program or a cinematic production. These clips are named "dailies" because at the end of every day the day's footage is collected and developed, synchronized to the audio channels, and prepared for viewing. Dailies are viewed by members of the production crew, typically early in the morning before the day's filming starts, but sometimes during a lunch break or at the end of the day. It is common for several members of the production team including the director, cinematographer, editor, and others to view and discuss the dailies. Dailies are sometimes separately reviewed by producers or executives who are not directly involved in day-to-day production to ensure their investment is on track as expected. Sometimes multiple copies of the dailies are distributed to for individual viewing, via a secure Internet connection or via physical media such as a DVD.

Dailies indicate how the overall filming and actors' performances are advancing. At the same time, in industry jargon the term can refer to any raw footage, regardless of the actual date of capture. In the UK and Canada, dailies are called "rushes" or daily rushes, referring to the speed required to quickly turn-around the print for viewing. In animation projects, dailies are called rushes and you may hear the dailies review called "sweat box" sessions.

Active monitoring of dailies allows the film crew to review images and audio that were captured the previous day, and it provides technical as well as artistic analysis of the captured media. Technical problems can be caught and resolved quickly. Directors can evaluate and modify the actors' performances as well as adjust camera angles and scene positioning. If a scene must be reshot, it is best to address the need immediately rather than later in the process when sets may have been torn down and actors have left the production schedule.

Realistically the process of reviewing a dailies sequence is monotonous, as dailies often include multiple recordings of the same scene with minor changes or adjustments. High Definition digital video dailies can be as big as 2K resolution (2048×858, 2.39:1 aspect). Many productions have a main production unit which does all primary cinematography and one or more smaller teams shooting additional "pickup" shots, stunts, locations, or special effects shots. This additional video is included with the main unit footage on the dailies reels. Because of the way the dailies are processed, when a unit shoots with more than one camera typically all the shots from one "A" camera will be followed by all "B" camera shots, then by the "C" camera and so on. Wary of the time required to review every shot, ordinarily only a small amount of the previous day's footage is viewed; and with digital media, footage can be fast-forwarded as desired. Sound that was recorded without simultaneous picture recording is called "wild sound" and is sometime included in the dailies. Visual effects shots are collected daily for viewing by a visual effects supervisor. These dailies contain the previous day's work by animators and effects artists in various state of completion. Once animation or character generation requires additional feedback from the director, the supervisor will collect specific dailies and screen for the director either as part of the normal dailies process or in a separate visual effects dailies screening.

As most modern editing is accomplished on computer-based non-linear editing systems, keycode numbers are logged on the media which assign a number to each frame of film and are later used to assemble the original film to conform to the edit. Dailies delivered to the editing department are essentially "proxies" and already have timecode and keycode numbers overlaid on the image. These reference numbers assist the later assembly of the original high-quality film and audio to conform to the edits.

Modern video cameras can record the image and sound simultaneously to video tape or hard disk in a format that can be immediately viewed on a monitor, eliminating the need to undergo a conversion to create dailies. Audio synchronization can be very important, and the film methods of using clapperboards and manual adjustments are still found on production sets. Sound collection and synchronization need to be done for every take.

Rushes and dailies can be used to create trailers and "sizzles," short promotional clips, even if they may contain footage that is not in the final production.

SOURCES AND TYPES OF METADATA

Metadata, or data about data, is sourced at many different stages in the production process and supplied from many different foundations. Broadly we can separate the sources into machine- versus human-generated metadata. Types of machine-generated metadata include but are not limited to:

- Camera technical specifications like field of view, sample rates, etc. and captures of the specific settings such as recording format, output format, etc.
- Technical lighting settings including any color grading information and equipment settings
- Technical audio settings and equipment settings
- Ancillary equipment settings such as time code generation
- Metadata collected in the dailies creation process or in the dailies transmission process
- Metadata captured by asset management systems during ingest including technical file data such as wrapper and media formats, Sidecar XML descriptive data, captioning data, etc.
- Metadata created in the editing process, including edit mark-in/mark-out points, multiple media file selections, and specific edit tool project files
- Metadata generated in the effect creation or animation processes including insert mark-in/mark-out points and specific tool project files
- Metadata created in a compression or transcode process

Types of human-generated metadata include but again are not limited to:

- Title, date, and location
- Light, weather, and time of day references
- Production schedule references
- Tracking numbers
- DIT notes
- Description of scenes
- Credits including director, actors, cinematographer
- Notes from dailies review
- Scripts and script notes
- File annotations to note points of interest in the digital files

There are many sources for external metadata which can be used to augment a particular asset. These include the Entertainment Identifier Registry numbers (EIDR), Ad-ID identifiers for commercial advertisement and rights tracking purposes, company-specific media identifiers like house numbers, Universal Unique IDentifiers (UUID), metadata sourced from external descriptive libraries, and work orders or requests for media access, delivery, or edits.

Machine Learning (AI) applications have become a major source for metadata creation. Using Asset management systems as a repository, machine learning software can be used to perform metadata augmentation and annotation to digital video and audio media. The output of these systems is typically an XML (eXtensible Markup Language) or JSON (JavaScript Object Notation) file that frame accurately documents the media or its proxy. Cloud-based, consumer-driven systems like those offered by Amazon Web Services and Google can provide well-trained software systems for generic applications such as speech to text, scene change, or celebrity recognition. More broadcast and media focus platforms such as Veritone, Graymeta, and Zorroa provide focused engines with particular characteristics to manage specialized jobs like content recognition for compliance (nudity, prohibited language, prohibited actions like smoking, etc.), location of product and logo recognition, object recognition, etc.

THE ROLE OF THE DIT

As digitization becomes more important in our production chain, more tasks concerning data management have evolved and increased, and the position of the Digital Imaging Technician (DIT) has been created to address this ever-growing need. The DIT position was originally created to manage the transition from the long-established film movie camera medium into the current digital cinema age. As a camera department crew member, the DIT works in collaboration with the cinematographer on operational workflows, system integration, the production's camera settings, overall signal integrity, and any image manipulation and effects to achieve the creative vision of the director and the cinematographer in digitized media.

Involved in the entire end-to-end production, DITs are responsible for preparation, on-set tasks and post-production and, in fact the DIT connects the on-set work with the post-production operations. The DIT role has become a blend of several other positions, such as Video Controller, Video Shader, or Video Engineer. DITs support camera teams with technical and creative digital camera management to guarantee the highest technical quality as well as the safety and security of the digital files. They are the established responsible party for managing all data on set, including system quality checks and reliable file backups. Once a project enters post-production, the DIT manages the delivery of recordings to the post-production team, including quality control and generating working copies if they are required. Typically, the DIT must ensure that original camera data and all associated metadata is backed up at least twice daily, ensuring data integrity with checksum verification. Important productions require all backups to be made on LTO tapes, a more sturdy and reliable method of storage compared to camera hard drives and spinning disk storage.

The DIT's role on-set has become important in assisting cinematographers, who are conditioned to work with film stock, in achieving their artistic vision through digital tools. Through monitoring exposure, building Color Decision List (CDL) and "look up tables" (LUTs) on a daily basis for the post-production team, the DIT actively assists the team. Across all equipment on the set, the DIT manages settings in digital camera's menu system for recording format and output specifications. The DIT is responsible for collecting and securing any digital audio recorded by an external recorder operated by the Production Sound Mixer.

Often working beside the DIT, the data wrangler position was created as a support role for managing, transferring, and securing all the digital data acquired on-set via the digital cameras and depending on the scale of the project, the DIT can cover the data wrangler position but usually not the reverse.

THE ROLE OF THE DATA WRANGLER

Data wrangling is the process of mapping and / or transforming data from one "raw" form into a different format to add value for a variety of purposes such as interoperability, analytics, and to meet archive standards. The process is sometimes called data "munging." A data wrangler is the person who manages and performs these transformation operations. This may include data manipulation, aggregation, visualization, training a statistical model or machine learning (AI) software, and many other processes. As a process, wrangling typically follows a set of general steps which start with data extraction in a raw form from the data source, "munging," or sorting the raw data using algorithms or parsing the data into predefined data structures and depositing the resulting content into a database for storage and future use. This munged data comprises one important form of metadata for our media purposes.

Data wranglers can be found in most industries today, not just media and entertainment but the position can hold considerable importance as the coordinator for acquisition of data across many different sources and devices – in our Internet of Things (IOT) connected world, this has become increasingly important. Specific duties mirror those of a storage administrator working with large amounts of data, involving both data transfer from instruments to storage grid or facility as well as data manipulation for re-analysis via high performance computing instruments or access via Internet infrastructure-based and AI tools.

Our data wrangler's focus is on the massive amount of data generated by cameras, digital audio recorders, and productions and editing tools. When data transformations are required, for example between different camera systems, these include actions such as extractions, parsing, joining, standardizing, augmenting, cleansing, consolidating, and filtering to create desired wrangling outputs that can be leveraged in the downstream post-production processes. Depending on the amount and format of the incoming data, data wrangling has traditionally been performed on spreadsheets or via hand-written scripts in languages such as Python or SQL. Sometimes data wrangling can be managed in asset management systems prepared with the proper data model and conversion tools.

EXTENSIBLE MARKUP LANGUAGE (XML)

Extensible Markup Language (XML) is a computer code language that defines a set of format rules for encoding documents that are both human and machine readable. The free and open standards defined in the XML 1.0 Specification that was adopted by the World Wide Web Consortium (W3C) as well as several related specifications define XML. The primary application goals of XML emphasize simplicity, generality, and usability across the Internet. It is a textual data format designed to support Unicode for adoption by different human languages. Although designed to manage documents, XML is widely used for the representation of arbitrary data structures such as those used in digital media metadata structures. Several schema systems aid in the definition of XML-based languages, and programmers have developed many application programming interfaces (APIs) to aid the processing of XML data between tools and systems.

XML was developed as a specification to address the proliferation of non-interoperable document encoding structures. Today, hundreds of document formats using the XML syntax have been developed, including SOAP, SVG, and XHTML. XML-based formats have become the default for many office-productivity tools and the rich features of the XML schema specification have provided the base language for communication protocols. XML has come into common use for the interchange of data over the Internet. It is common for a specification to comprise several thousand pages as many of these standards are complex. XML is widely used in a Services Oriented Architecture (SOA) where disparate systems communicate with each other by exchanging XML messages. The message exchange format is standardized as an XML schema (XSD). This is a typical method used by media software systems for the exchange of relevant metadata or command sets.

KEY XML TERMINOLOGY

When looking at an XML file, understanding some terms and usage will help you decipher the meaning of the document. The material in this section is based on the XML Specification. This is not an exhaustive list of all the constructs that appear in XML in any way – it is only provided as an introduction to the key concepts encountered in media system and metadata usage.

Character: An XML document is a string of characters. Almost every legal Unicode character may appear in an XML document.

Processor and application: The processor analyzes the markup and passes structured information to an application. The specification places requirements on what an XML processor must do and not do. The actual application is outside its scope. The specification calls it a "processor," but coders typically to it as an "XML parser."

Markup and content: The characters making up an XML document are divided into markup and content, which may be distinguished by the application of fairly straight-forward syntactic rules. Strings that constitute markup either begin with the character < and end with a >, or they begin with the character & and end with a;. Strings of characters that are not markup are content. It is important to note that whitespace before and after the outermost element is classified as markup.

Tag: A tag is a markup construct that begins with < and ends with >. Tags come in three flavors:

- start-tag, such as **<section>**;
- end-tag, such as **</section>**;
- empty-element tag, such as **<line-break />**.

Element: An element is a logical document component that either begins with a start-tag and ends with a matching end-tag or simply consists of an empty-element tag. The characters between the start-tag and end-tag, if any, are the element's content, and may contain markup, including other elements, which are called child elements. An example is **<greeting>Hello, world!</greeting>**. Another is **<line-break/>**.

Attribute: An attribute is a markup construct consisting of a name–value pair that exists within a start-tag or empty-element tag. An example is **<img src="madonna.jpg" alt="Madonna" />**, where the names of the attributes are "src" and "alt," and their values are "madonna.jpg" and "Madonna" respectively. Another example is **<step number="3">Connect A to B.</step>**, where the name of the attribute is "number" and its value is "3." An XML attribute can only have a single value and each attribute can appear at most once on each element. In the common situation where a list of multiple values is desired, this must be done by encoding the list into a well-formed XML attribute with some format beyond what XML defines itself. Usually this is either a comma or semi-colon delimited list or, if the individual values are known not to contain spaces, a space-delimited list can be used. **<div class="inner greeting-box">Welcome!</div>**, where the attribute "class" has both the value "inner greeting-box" and also indicates the two CSS class names "inner" and "greeting-box."

XML declaration: XML documents may begin with an XML declaration that describes some information about themselves. An example is **<?xml version="1.0" encoding="UTF-8"?>**.

With a little basic knowledge and by carefully reviewing an XML document, one can usually discern the meaning of the file without a coding background.

CLOUD IMPACTS

The ubiquity of cloud infrastructure has simplified the access of production teams to dailies, and enabled the DIT and Data Wranglers a faster, more comprehensive gateway to tools, storage, and automated backup processes. As more metadata is generated and more munging is required, cloud tools provide a trusted platform with fast ingress and egress. Competition in the cloud services market has begun the inevitable process of driving cloud infrastructure to more affordable levels, and software manufactures have begun to leverage cloud native designs with microservices so that a DIT can customize their platform to a production's unique requirements. Storage options and automatic disaster recovery options for data are a welcome support for the DIT's protection measures, and the advances in security have made these platforms safe and reliable.

CHAPTER 4

Post-Production

Section Editor: Arjun Ramamurthy

INTRODUCTION

Wikipedia notes that "*Post-production includes all stages of production occurring after shooting or recording individual program segments.*" Expanding on this understanding, this chapter will walk the reader through the processes that occur after initial (or principal) photography and on-set sound recording. We will examine motion picture and television together, detailing how content is brought or ingested, into the post-production process and how the content is finished, mastered, and versioned.

Physical production is typically limited to acquiring the sound and picture plus references for visual effects.

Post-Production encompasses everything that happens *after* that, all the way through to the finished product (Figure 4.1).

Traditionally, post-production includes creation of the dailies, editorial, visual effects, non-production sound (which includes design, effects, foley, music, dialogue re-recording, and mixing), color correction, and finishing of the content. In broad strokes, the middle area below is the focus of this chapter:

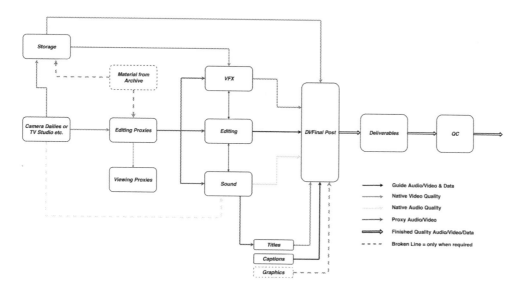

FIGURE 4.1 Simplified Generic Post-Production Overview (Movies and TV)

POST-PRODUCTION PROCESS for THEATRICAL PRODUCTIONS

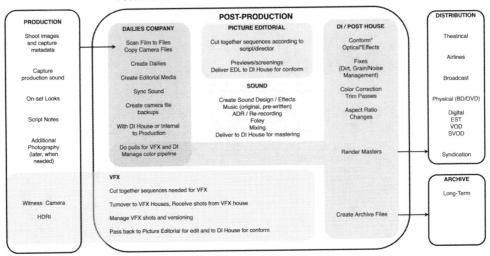

FIGURE 4.2 Post-Production within the Supply Chain

The careful management and curation of content is critical for all post-production processes, so that it can be edited, graded, scored, mixed, and prepared for the various distribution channels, which concludes with addressing localization issues such as dubbing and subtitling as well as handling regional edits based on localized concerns and formatting for platforms and devices. In addition to detailing the post-production workflow, this chapter will also introduce the reader to the different skills and job descriptions tasked with taking a raw product and turning it into the final dish – sticking with the traditional saying – "**We will fix it in Post!**"

Traditionally, the tasks during the post-production processes were clearly delineated with little cross over between the roles shown as different colored boxes in Figure 4.2. However, in today's digital post-production "lab," we often see that personnel typically work across traditional lines, and that the post-production process can now be described as an eclectic mix of storytelling, technology, and logistics management. With the introduction of more and more digital tools, from digital cameras, digital sound management processes right through to digital editing and image manipulation tools, it has become possible to operate post-production departments that range from simple, self-operated, laptop-based editing of sound and pictures that are often used for the end-to-end processing of simple television programs, all the way up to complex logistical operations that engage multiple facilities that would be used to post-produce a major motion picture or episodic television series. This has led to a rapid broadening of the list of skills and job descriptions required in current post-production areas.

Another way to understand post-production is to look at an overview of the key roles and skills of modern post-production. While the exact route through post depends on the complexity and budget of a program, one way to look at post-production workflows is to think about the skills needed. Some of the key areas to consider are (Table 4.1):

While each area detailed above has the requirement for a specific set of skills, all are closely related to the whole post process. On larger productions, each area often has a Producer or Supervisor whose role is to coordinate the workflow. On smaller productions a single person can often manage the end-to-end post process. In general, the organization chart then is similar to the one shown below:

TABLE 4.1 Post-Production Areas and Roles

Area	Role/Task (where the key roles are based)
Studio/production	Oversight of budget, schedule, crew and vendors, and broad vision for the show
Dailies	The processing of camera footage after shoots or "dailies/rushes" to prepare for viewing and editing
Editorial	Storytelling! Cutting and editing together what will become the final story, picture, and sound
Visual effects (VFX)	The enhancement or creation of images that could not necessarily be realized on set
Sound	The production and enhancement of the audio tracks including music and special audio effects
DI/final post	All of the elements come together for the creation of final versions of the program or movie at the highest quality

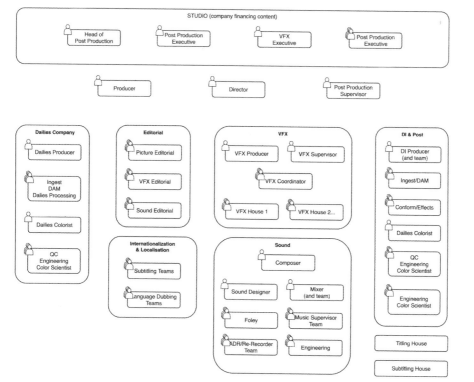

FIGURE 4.3 Post-Production Roles

Large and complex productions are usually arranged so that each department is organized under a supervisor and a producer. It is the role of the department Supervisor to deliver to the creative vision of the Director and the role of the department Producer to deliver within the budgetary and scheduling constraints imposed by the TV or movie Producer.

An area that clearly demonstrates each of these roles is Visual Effects (VFX), especially on large productions. The VFX Supervisor and the VFX Producer manage numerous VFX vendors that are tasked with delivering a variety of interrelated shots. The VFX Supervisor will regularly look at the images from a technical and creative aspect, while closely working with the VFX Producer to help determine and resolve issues that may affect the overall VFX budget and schedule. Additionally, these Supervisor and Producer responsibilities generally encompass managing numerous VFX vendors that are tasked with delivering a variety of shots. In turn, each VFX vendor also has an in-house VFX Producer and VFX Supervisor, who similarly mirror the organization of the production's VFX Supervisor and Producer. Smaller productions, especially those that use one VFX vendor, often will use the VFX vendor's in-house VFX Supervisor and VFX Producer to oversee the VFX process.

During complex post-production, the timing of each process is critical so no one area is held up due to a previous process overrunning or not delivering what is expected. It is the responsibility of the Post Producer or Supervisor to ensure each area or department delivers on time and that the groups are working together to the overall production timeline. The Post Producer or Supervisor's role is to ensure the smooth on-time transfer of content between departments and to ensure that each department is operating as efficient as possible.

This is necessary not just to meet the deadlines for the delivery and transmission of TV programs or the launch and premiere dates for movies, but critically to ensure internal deadlines during the post process itself are met. For example, the delivery of proxies for viewing for a specific location or computer graphics (CG) delivery deadlines or music composition deadlines, etc. In a complex production, the list can be incredibly long involving many facilities, artists, and brute force processing – often referred to as "rendering" (which can become a byword for unexpected delays)!

It has to be remembered that the key process in post-production is not technical and can often cause reorganization processes. It is inevitable that there will be numerous edits and re-edits of the content as the Director focuses on the pacing and final timing of shots and performances. During this iterative process, the entire program will be shown to audiences as test screenings often known as "Friends and Family" or "Preview" screenings to judge whether the story line is playing out as intended. It is important to realize that the dialogue, music, effects, as well as visual effects and drama shots should be in a state that can be shown to the "preview" audience who may not understand that the preview is not the final product. It falls to the post-production Supervisor to coordinate all of these departments to meet each deadline.

At a high level, the post-production timeline and hand-off between departments can be visualized as Figure 4.3. As stated earlier, the reader should keep in mind that while these tasks have been delineated, for purposes of clarity today's multifaceted tool sets permit a very small group of individuals on highly mobile laptops to execute most, if not all, aspects of the post-production pipeline. This has enabled faster delivery and also delivery to numerous additional channels and endpoints, as explained in succeeding chapters.

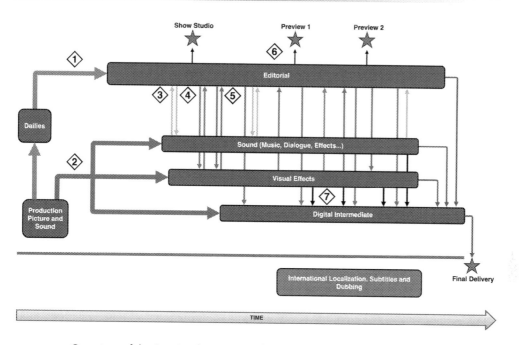

FIGURE 4.4 Overview of the Post-Production Workflow and Interactions between Groups

Figure 4.4 provides a generalized overview of the Post-production workflow for a typical motion picture. In step one (1), Dailies (see Dailies) are transferred to Editorial (see Editorial) which is the hub of the production. Editorial will go through multiple iterations until the cut becomes "locked" with the final timing of shots and performances. Often times, the editing is rearranged, or re-shoots are required, or additional photography is needed, repeating the production process again in order to capture missing or additional shots or performances needed to tell the story as desired by the creative team. This final locked cut is the version utilized through finishing.

While the Editorial team is revising their version, the Sound and Visual Effects teams access the production assets, and based on an initial shot compilation from Editorial begin to work their respective areas, as outlined in steps 2–4. As the Visual Effects and Sound teams generate temporary versions, these are sent back to editorial, so that they may be "cut-in" into the timeline to be viewed in continuity. This can go through multiple iterations with multiple versions being transferred back and forth, as denoted by the yellow arrows to Sound and the red arrows to Visual Effects. Simultaneously, the Editorial department (Step 6) is required to generate the "Preview" and other review versions to gauge audience responses and to revise the story in response to the feedback.

As the Post-production process proceeds, the Digital Intermediate team is brought (Step 7) on to start working with the production assets and develop the overall look of the movie. Once again, this is an iterative process, taking in intermediate versions of the cut from Editorial, as well as temporary visual effects and temporary audio. As each department converges towards the final versions, they are all compiled and coalesced into the final delivery.

At the same time the localization teams (Step 8), usually under the international group, are working on dubbing the feature into many different languages, developing subtitles for the different regions and possibly incorporating special regional or territorial material to meet cultural needs.

POST-PRODUCTION IN DETAIL

It is worth spending a bit of time looking in more detail at each of the post-production processes. Many of these processes apply to both movie and TV post-production although the scale may vary considerably; however, it is important to understand the background no matter if post is a single person self-operating or a major multi-company movie!

Preparation

The goal of post-production is to ensure consistent and accurate representation of the Director's and Cinematographer's creative intent, from Principal Photography and Dailies through Post-production and Delivery of the motion picture or TV program. As the final deliverable platforms continue to expand, it becomes more and more important to identify which technology options affect each part of the post-production process.

No matter how simple or complex the post-production process, planning is the key to success; and in order to achieve this process as smoothly as possible, planning before production begins is always time well spent. There are logistical considerations such as where departments such as Editorial or VFX will be located, where they will be sourcing their gear and who will supply support for the gear.

Do not forget that special consideration should be given to establishing clear and documented content security guidelines for all members of production before production begins. It's very difficult to institute new controls and safeguards once production has begun – and sometimes too late if content has already been leaked, transferred via insecure public Internet or simply storyboards left in photocopiers!

Technical Preparation

Every department should know the exact requirements for each process and how they will receive and pass on material. There are numerous decisions that must be made; the wrong choices adversely impact the final deliverables. Here are a few of the technical requirement questions that should be asked and clearly documented before post-production planning starts:

Resolution

a.   What is the maximum final output resolution? – There may be several targeted at different outputs but the highest sets the bar.
b.   What is the working resolution for VFX? – It should be good enough for early Previews and viewings but is fast enough to work with during the Editorial processes.
c.   What Aspect Ratio will be used, and will it be protected? – Content has to be formatted for many displays from full blown cinema to mobile devices.
d.   What are the different camera acquisition resolution(s)? – Not all cameras are the same, especially if specialist or mini-cams are used.

e. Is there a common DI resolution? – This is the target! Even for TV there may be a common mastering resolution.

f. What cropping or padding of the image will be needed for visual effects and/or stereoscopic 3D? – Image stabilization or 3D image alignment zooms images; this can affect framing and quality if not known.

g. Is there a standardized debayering toolset? – Vital for single sensor electronic cameras. Poor debayering can wreck quality!

Figure 4.5 illustrates a sample resolution diagram that would be generated for a feature production, calls out the different cameras used on the show, and the different acquisition resolutions, and how they flow through the post-production chain. It also shows the common color space (ARRI Wide Gamut) for the show.

Color Pipeline

Note: For TV workflows, the color pipeline is fairly standardized, however some of the questions below are extremely important to ask before Post-production can begin on a Motion Picture in order to determine the color pipeline and workflow.

h. Who will be in charge of the "Look" for the dailies color and what display output is being targeted? Often, the DP sets the look, but others can as well. Typically, a Rec.709 display is used as the target display.

i. Who is generating the Look-Up-Tables (LUTs) for the look? Are they the clearing house for the LUTs in the future (to give to VFX houses and marketing vendors)? LUTs are usually provided by the DI facility, but sometimes they are provided by the Dailies facility, the DIT, or DP.

j. How will Look-Up-Tables (LUT) and ASC CDLs be managed?

k. What is the "Working Space" for the dailies grade? Will it be the same for the final DI grade? This is typically either a camera proprietary working space or an ACES working space (Academy Color Encoding System).

l. What is the creative "White Point"? Typical choices include D65 and D60.

m. If required, how is film being scanned?

n. What is the bit depth and file format of the image files that will be used for VFX exchange and DI? Typical choices include 16-bit EXR, 16-bit DPX, or 10-bit DPX.

o. What will be the target display for the final DI hero grade? How should VFX houses be viewing the shots? The typical choice for DI is a P3 D-Cinema projector, but VFX houses may view in either Rec.709 or P3, depending on their resources.

p. Will a neutral or balance grade be needed for the VFX plates?

q. What Color Spaces need to be supported for display of the content? It is important to know what color output standards are required such as Rec 709 for HDTV, DCI-P3 for theatrical, Rec 2100 for HDR …

Key Deliverables

r. Standard Dynamic Range only? Requires a simpler workflow with fewer options. Images need no dynamic metadata but are limited in dynamic range.

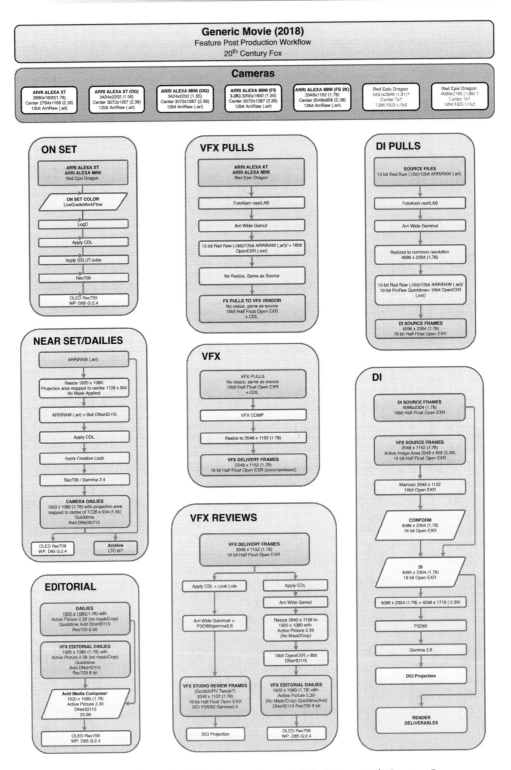

FIGURE 4.5 Sample Resolution Workflow from a Generic Title @ Twentieth Century Fox. Workflow Courtesy of Twentieth Century Fox

s. Are High Dynamic Range versions needed? There are two primary production standards (see HDR section) but for HDR content, it is vital everyone knows the target brightness and the "reference white level" needed.

t. Final Resolution for Theatrical.

u. Final Resolution for Home Entertainment Deliverables?

v. UHD for TV programs? UHD TV programs are not actually "4K," it is 3840 pixels wide and 2160 pixels tall, so it's very important to read the delivery document during planning. HDTV and SDTV deliveries are relatively easy!

Figure 4.6 shows the product of a "Deliverables" discussion for a production, highlighting the different versions, how they will be rendered, at what resolution, what the color space is and the delivery formats for each deliverable.

Audio

w. Is the production going deliver in an immersive sound format, e.g., Dolby Atmos™? It is very important to know the primary listening "environment." This will impact mixing, QC, and Preview processing as well as the final finished output.

x. Is the delivery a fixed channel-based option, e.g., Stereo, 5.1 or 7.1? Easier to understand during workflow planning as setup and listening environments are well documented.

y. Are multiple audio options needed? This is the sound department's nightmare. No one can satisfy every audio reproduction option, but during planning the target options must be documented.

z. Is down-mix metadata required? This is very important for TV where only one audio option is sent. It is usual for the viewers' TV to downmix 5.1 to stereo using the supplied metadata.

aa. What plugins will be used within NLE and audio tool (e.g., ProTools®) and are they supported on all systems? What you don't want is a holdup when the sound is not right after moving from one facility to another – even worse, no one will notice until the Director is at a Preview!

Editorial Considerations

bb. What is the Codec and Bitrate for the Editorial processes? Too low and errors or issues with content may be missed – too high and processing takes too long.

cc. Will playback be limited to stereo or will multichannel be needed? Not usually an issue but a minefield if Editorial requires it mid-post.

dd. Is remote Editorial Review needed? If yes, then secure fast systems need to be in place that are easy enough for the recipient to use on location or some other method of remote review will need to be setup.

Archival Considerations

These will vary Studio by Studio, Broadcaster by Broadcaster and Production by Production but the key questions are:

ee. Will the camera originals need to be archived?

ff. Will LUTs and camera settings need archiving?

gg. What level of audio archive is needed?

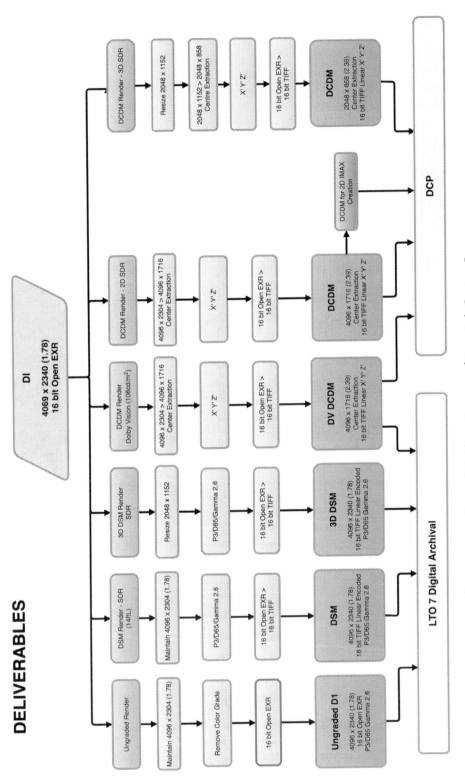

FIGURE 4.6 Deliverable on a "Sample" Feature Production Workflow Courtesy of Twentieth Century Fox

    i.   All elements – free from any processing?

    ii.   Processed Stems?

    iii. Pre-mix Foley?

    iv. Submix masters?

    v.   Main and M&E only? (often for simple TV programs)

hh. Components Archive

    i.   Ungraded file archives

    ii.   Graded File archives (with or without the looks rendered in?)

    iii. DCDM archive

    iv. Editorial archive

Finally, all of these considerations will inevitably be a balance between the post-production schedule, the post-production budget, and the final requirements from Editorial.

The post-production Supervisors and Producers will ultimately determine how the post-production budgets will be best used and will track the spend to make sure costs stay under control.

POST-PRODUCTION SPECIFIC FOR TV

It is fair to say, TV content covers a vast range of program budgets! This means the post-production process can range from the director cutting sound and vision content on a laptop through international TV episodics that have post processes which are more complex than movies – usually due to the sheer number of episodes and overlapping seasons.

For many daytime or one-off programs, preparing an edit for delivering to television for traditional broadcast can be deceptively simple:

- Import the camera files
- Import additional content (downloaded music, etc.)
- Cut the sound and vision content
- Color correct
- Add burned-in graphics or text (e.g., lower third names or end credits)
- Mix the sound
- Export the completed program

Simple as it seems, there is a very high chance the output will fail the broadcaster or distributor's QC process or just be rejected even before QC!

As with all post-production, it is vital for anyone – from a self-operated editor to a post-production Supervisor on a large episodic – to read and know the broadcaster or distributor's delivery guidelines. To be fair, these are not really guidelines but are more of an instruction set that must be followed to deliver a TV program for transmission or distribution.

The delivery guidelines which must be followed precisely should include information on:

- What video format has the commissioner asked for? This could be UHD, HD, or even SD.
- Which frame rates are acceptable? Remember fractional and interlace are often the only options for TV programs.

- What codec does the broadcaster require? Get this wrong and the program won't even get through to QC – automated "gateways" will simply bounce the file.
- What is the audio channel layout? A program could get a long way through the chain before this is noticed – maybe even to transmission!
- What is needed for the top and tail, line-up, slate, black, and silence, etc.? Nothing is worse than a season being rejected because the slate is incorrect!
- What duration and commercial break parting is needed? As with the above, nothing is worse than a whole season being rejected because the commercial black is a second too short!
- What audio volume or loudness standards are needed? This is not negotiable in many places. TV is regulated and non-compliant programs are rejected, or even worse, they are automatically "normalized" by the distribution path.
- What metadata and "paperwork" (pdf or word) information must be sent with the program? This is used to confirm the program details. Remember a broadcaster will be receiving tens or even hundreds of programs a day. Whether metadata or paper, the details must be accurate and clear, or a program could just get lost.

Also, some TV content is commissioned for Over-the-Top (OTT) as well as traditional distribution. Sometimes, these are identical versions, in which case the same file is used but often the content layout is different (no commercial breaks or no teasers) which may mean additional material is needed.

THE END GAME!

The process and requirements can diverge dramatically, depending on the target platform(s) for the finished content (e.g., Cinema, TV, mobile, OTT, etc.) and all can have different and sometimes mutually exclusive requirements. It is very important that the overall post Supervisor has a solid grasp of the technical and editorial requirements for each version.

The following sections work through the post-production processes that might be required to deliver the final deliverable(s); however, one thing to remember is ALL of the facilities used during "the end game" must have correctly aligned equipment that follows industry standards, including and not limited to:

1   Displays
2   Viewing rooms
3   Audio reproduction equipment
4   Reference line-up capability
5   Fully trained operations/engineering staff
6   The right talent or the ability to bring on artists based on production needs (e.g., colorists)

DAILIES IN MORE DEPTH

Once production starts on location or at the studio, the content (either on film or on digital camera files) is shot and will need to be viewed daily. As outlined earlier, these are known as "dailies" because usually they arrive at the end of each day's shoot. The dailies process prepares the images, synchronizes the sound to the picture, and organizes the files for review by the production crew after each day's shoot. In some regions such as the UK and Canada, dailies are usually referred to as "Rushes" or daily rushes, referring to the speed at which the film prints were originally developed back when film cameras were used.

Dailies give the Director and production crew an idea of how the filming and actors' performances are progressing. If the Director sees any issues, he or she can organize retakes before the production changes location, or the talent is dismissed or "stood down."

The dailies process can be carried out at a dailies facility or near-set by either the Digital Imaging Technician (DIT) or the crew from a dailies facility. The DIT is typically a camera department crew member who works in collaboration with the cinematographer on work-flow, systemization, camera settings, signal integrity, and image manipulation to achieve the highest image quality and creative goals of cinematography.

If the production is shot on film, the dailies house will work with a lab to process the film and transfer it to files. This process typically involves the development of the negative film and then digitizing using a film scanner in high speed, and lower resolution – often referred to as a telecine. Once the film is developed and scanned, the raw camera images are treated the same as with digital camera footage (as described below) in order to create the dailies color, synchronize production audio, and back up the camera footage.

If the production is shot on digital camera, the DIT or dailies house will offload the camera files from the camera "mags" to their storage. Camera RAW formats have to go through a debayering process to convert the images to RGB in order to be processed for dailies, while compressed camera formats like Apple ProRes™ maybe left as received, depending on the editorial/screening workflow, or the files can be transcoded to more compressed file formats for dailies viewing.

In addition to offloading the camera files and debayering, the DIT or dailies house is responsible for:

1 Checking the media contents for possible issues (duplicate time code or file names).
2 Collecting and backing up copies from of the original camera footage and audio recordings.
3 Provide checksum-verification of the copied data ensuring that copies are identical to the original footage.
4 Complete shot reports (scene/take information, filters used, comments, etc.) based on script notes.

For simple TV programs the Dailies process can be as simple as copying the camera card files to one or more (as a backup) hard drives followed by direct viewing on a laptop or even a hotel TV! It is also usual to wipe (delete the contents) the camera cards after copying so they are ready for re-use the following day. If this is the case, it is more than vital the copy is viewed end-to-end before the camera card is wiped (for obvious reasons). The same is true for motion pictures, although due to the size of the camera files, it usually takes longer than a day to return back the camera cards. Also, data verifications are also done before wiping the camera cards to make sure the integrity of the camera files is intact after the copy.

What is really important for any Dailies process is speed! The procedure is a tradeoff between getting the content in front of the director and the maximum quality that can be achieved at speed. For TV, this is why the process is typically repeated when shots have to be delivered to Visual Effects or in Digital Intermediate or when needed for other final finishing needs, such as trailers, etc. where full quality copies are essential. For motion pictures, camera footage is shot as RAW files which are debayered and transcoded into a certain bitrate of files for Editorial which are also used to screen high-quality dailies on a projector. The RAW files are also transcoded into even lower bitrate files to use on dailies systems for viewing on iPads™ and computers. The bitrate of the editorial files and dailies system files is usually discussed before production starts so that everyone is on the same page.

As the speed of film and file processing have increased and the time needed for file copying and backup have dramatically decreased, some post-production houses now offer the choice of a single digital scan that can be used for the dailies process as well as the final digital intermediate process. This means the easy-to-damage film negatives are only touched once, and the post-production house maintains the digital scans for downstream use. But there may be trade-offs. For example, the original negative's dynamic range may be reduced by only flashing the negative once, rather than twice to capture the entire dynamic range of the negative. The same is true of debayering digital camera images, where a simple and fast reconstruction process would be used for Dailies, while a more accurate and time-consuming process would be used for final, higher quality images. Typically, this choice, as many others in the post-production process, would be driven by the needs determined in the post-production preparation period where the end deliverables are clearly identified, and the budget constraints are imposed.

The dailies house will also load all production sound onto their storage in order to synchronize the sound to the image and will create the media files used by editorial and for all of the other dailies systems including projectors (in the case of theatrical feature workflows), television monitors, and mobile devices such as iPads™.

Still, the most effective and useful way to sync audio is a clapperboard. It not only pinpoints the sync point but also provides a visual marker with the audio identifier for the take and as a byproduct, alerts all on-set that sound and vision are recording. The clapperboard is labeled to identify the scene, shot, and take number for the camera. The numbers are also read aloud to label the audio recording. Once camera and sound are rolling, a camera assistant will close the clapper creating a visual and auditory reference point.

During the syncing process the dailies technician will look at the numbers on the slate board on the camera file (or on the digitized images of film) and then match the numbers with the verbal slate. Then, the technician looks for the frame where the clapper first closes and for the beep or clapping sound on the audio tape, adjusting one or the other until they happen simultaneously when played back. The operator is assisted in this with automated systems that permit automatic syncing of picture and audio using the synchronized timecode on the camera footage and the audio recording.

The dailies house is also responsible for the backup of all camera and sound files for safekeeping and disaster recovery, usually on LTO tapes.

The dailies company takes the reference looks set by the DIT and DP from set and creates a "dailies grade" or color for each of the shots for dailies viewing. The dailies color is meant to capture the color intent of the DP and Director and gives them a rough idea of what the shots could look like (otherwise, the original camera image might not represent exactly what the DP or Director intended). They may do simple (and sometimes complex) color correction to change the look of shots or to balance them within a scene.

Many productions will perform color correction based on the ASC CDL specification during dailies grading. This is a color decision list created by the American Society of Cinematographers. The ASC CDL allows for basic color correction information to be exchanged between different systems. The ASC CDL also enables color to be added as a separate step instead of rendering in the color, which gives more flexibility when more complex color timing happens later in the post process. The ASC CDLs can travel with the un-color-corrected camera files in order to create the dailies color non-destructively. However, for simplicity, the dailies company renders in the dailies color into the editorial and dailies viewing files so that CDLs do not have to be managed on these platforms.

To provide a specific example of a recent production where the DIT and crew acted as the dailies house, the following workflow was followed:

1　The production utilized the Alexa 65, and using Codex SXR drives the crew recorded full 12-bit 6.5K sensor data in ARRIRAW.

2　The 2TB capacity of each drive permitted approximately 43 minutes at 24fps.

3　The camera "mags," in this case the Codex SXR Capture Drives, once ejected from the cameras, were taken to the DIT station for downloading using a Codex Vault.

4　At the DIT station, the DIT used Pomfort LiveGrade driving BlackMagic Design's HDLink boxes.

5　The DIT applied a base 3D-LUT and performed a fine-tuning color grade using ASC CDLs.

6　The LUT used was designed by the Director of Photography (DP) and the DI colorist during post-production preparation.

7　The ASC CDL graded 1920×1080 4:2:2 signals were displayed for the DP on calibrated Sony 25" PVM OLED monitors in a light-controlled environment.

8　The data was then transferred to a mobile truck lab where a second Codex Vault unit took the 8TB Transfer Drives from set and transcoded the ARRIRAW files into 4K ProRes (as mezzanine files) for the fast creation of the editorial deliverables (DNX115) in FilmLight's Daylight system.

9　These editorial deliverables incorporated the color grade information from set, and also generated the ARRIRAW archives.

Once the Director views the dailies, he or she will make particular note of the "circle takes" or "selects" which are usually the shot takes that were noted on-set as the takes that had better performances from the actors. The Director will confirm which selects are good for use in Editorial. These selects are noted for future reference by Editorial.

EDITORIAL

Editorial is often thought of as one group that cuts together the story of the content, but actually several different functions occur which makes them the "hub" of everything in a production. In other words, nothing can function without Editorial! The **"Editor"** works with the raw footage transcoded by the dailies company, selecting shots and combines them into sequences which create a finished motion picture. However, supporting the "Editor" are the Picture Editorial, Sound Editorial, and VFX Editorial teams, with their supporting cast of assistant editors, coordinators, etc.

On its most fundamental level, film editing is the art, technique, and practice of assembling shots into a coherent sequence. The job of an editor is not simply to mechanically put pieces of a film together, cut off film slates, or edit dialogue scenes. A film editor must creatively work with the layers of images, story, dialogue, music, pacing, as well as the actors' performances to effectively "re-imagine" and even rewrite the film to craft a cohesive whole. Editors play an instrumental role in the making of the content and story.

There are several editing stages, and the editor's cut is the first. An editor's cut (sometimes referred to as the "Assembly edit" or "Rough cut") is normally the first pass. The film editor is involved through the pre-visualization and principal photography stages (aka Production) and in all likelihood is closely involved in reviewing dailies as production progresses. For the first pass the editor's cut might simply be a collection of the shots that represent the selects, in script order, which presents the first bed that the editor will continue to refine as shooting continues.

When all live action footage has been captured, the director can then turn his or her full attention to collaborating with the editor and further refining the cut of the film. This is the time that is set aside where the film editor's first cut is molded to fit the director's vision. In the United States, under the rules of the Directors Guild of America, directors receive a minimum of ten weeks after completion of principal photography to prepare their first cut. While collaborating on what is referred to as the "director's cut," the director and the editor go over the entire movie in great detail. Scenes and shots are re-ordered, removed, shortened, and otherwise tweaked. Often it is discovered that there are plot holes, missing shots or even missing segments which might require that new scenes be filmed. Because of this time working closely and collaborating – a period that is normally far longer and more intricately detailed than the entire preceding film production – many directors and editors form a unique artistic bond.

One thing to consider when planning the Editorial section of the post-production workflow is the potential impact of getting stuck in "The EDL" or "Edit Decision Loop" (Figure 4.7).

Once the TV program or movie enters the "EDL" there is a high risk it will never exit. This is where the Producer and Post-production Supervisor need to step in and provide all members of the creative team guidance with regard to schedule and budget.

Different Editorial Groups

There are several editorial groups working on content simultaneously during a production. These include Picture Editorial, Sound Editorial, and VFX Editorial. Each of these groups serves a specific function.

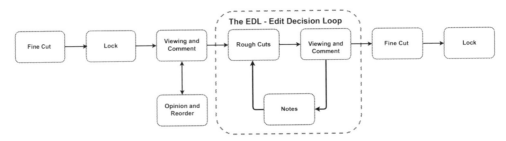

FIGURE 4.7 The Editorial Process – Edit Decision Loop!

- Picture Editorial is tasked with cutting together the story line of the content.
- Sound Editorial specifically works to get all of the sound pieces together for the content. See Section on Sound Editorial below.
- VFX Editorial puts together sequences that will be modified by a VFX house.

Picture and Sound Editorial are considered part of "Editorial" while VFX Editorial is part of the VFX team, but they all work together to get Picture Editorial assets to create the final version of the edit.

For a simple TV program, editorial is usually a small team consisting primarily of the Editor who brings in Sound (and VFX is needed) and possibly a color timer for a simple color grade.

For more complex productions, Picture Editorial consists of the Main Editor that edits the content and works closely with the Director to create the proper pacing of the content. The Main Editor usually has multiple assistant editors that help with organizing and ingesting the media as well as preparing the media in the editing system, exporting files from the edit and assisting the Main Editor as needed.

Sound Editorial and VFX Editorial also have similar structures and each group may have one or more coordinators to help with any other needs of the group like taking notes and scheduling sessions.

Marketing Editorial groups are not part of the production's editorial team, but they often use the same or similar media as Editorial in order to cut teasers, trailers, or promotional spots.

There are multiple editing systems available for use by editors and most editors prefer using a specific system. Currently, many editors either use Avid® Media Composer or Adobe Premiere, but some editors use Apple Final Cut X (with some still on Final Cut Pro 7). Blackmagic Design's Resolve is also starting to gain traction as an editing tool with editors.

As most productions want to work with specific editors, the system used on a show is usually dictated by the editor.

Typically, in the case of an Avid®-based editing system for a 24-frame production, Editorial is provided with MXF-wrapped files using the Avid® DNxHD 115 video codec and the audio files synced to the camera files in discrete day and shot (scene/take) order. These files are then imported in the editorial system as clips into bins and with the help of the AVID® ALE file, provide all of the relevant metadata information.

- As VFX completes their work on various shots, editorial receives a copy of the revised shots and replaces the original shots.
- As sound is mixed and created, editorial replaces their original scratch or production audio with the updated audio.

In the theatrical workflow, editorial results in a "Director's Cut" first which is written into most director's contracts. This is his/her first pass at the cut of the movie. The edit of the content usually gets whittled down from the Director's Cut and goes through several iterations until it becomes "locked" with the final timing of shots and performances which is the basis for the version that is seen in the end.

It is not uncommon for editing to be rearranged so that re-shoots or additional photography can be done (and repeating the pre-editorial production process again) in order to capture shots or performances that maybe missing or in order to tell the story appropriately based upon the audience's reaction to early versions of the movie. Once the cut is "locked," the timing and order of the shots will not change, so sound and VFX can really begin finalizing their work.

For some TV productions, the final version of the edit is what will be used as the basis of delivery. In these cases, color correction (see Digital Intermediate section) will be done on the editorial files and masters will be created. Often TV programs simply export the final version to the delivery format required by the broadcaster or distributor. See QC and Delivery for more detail. For theatrical and some episodic content, editorial creates an edit decision list or EDL that will be used by Digital Intermediate or DI post-production house for the conform, color correction, and finishing of the content (see DI/Final Post section).

VISUAL EFFECTS (VFX)

Many productions will use visual effects or VFX somewhere in the show, even if it is not a big budget movie. The VFX team works to either create images that could not be captured on camera or alter images that were captured on camera. This can range from things as simple as removing wrinkles from skin (also known as "vanity fixes") and removing wires and cabling on actors to creating completely computer-generated imagery or CGI characters and objects or extending set pieces to completely fabricate the full image from scratch for an alien world including the backgrounds.

Although many people typically associate VFX with big action movies, VFX can also be used to alter images. For example, a show may have gotten a great incentive financially to shoot in the wintertime when all of the trees were bare, but the movie requires the outdoor scenes to be in the summertime, so a VFX company may be hired to add leaves on all of the trees. One other area where VFX has become almost ubiquitous is with period dramas. People living in picturesque villages, especially in the UK, are no longer prepared to have roof antenna and street lighting removed for a shoot!

As mentioned earlier, each of the areas within Post-production benefit greatly from early preparation. In the case of VFX, especially for a big budget movie, planning starts in the pre-production stages, and a core VFX crew is tasked with collecting necessary data during the case of production. In some cases, the production may shoot footage such as Background Plates, primarily for VFX needs. So, although the majority of work for VFX happens in post-production, VFX planning typically starts in the pre-production and production phase. This preliminary work helps to streamline the VFX process during post-production and brings home the fact that as production processes have changed so have the traditional areas of post, pre and physical production.

VFX in Pre-Production

During pre-production for higher budget productions, pre-visualization, or "pre-vis," is created to help the filmmakers visualize the story with animations instead of only using storyboards. Pre-vis includes the conceptual artwork design done during pre-production, but incorporates the different camera moves, shot composition, and animations that would constitute the final shot. Once editorial begins, the production can use and update the pre-vis to post visualize ("post-vis") the live action shot with temporary animations and temporary visual effects to fill in the story editing while the VFX is being finalized.

Shots that are "heavy" with VFX can take weeks or months to complete, so in order to allow Directors to see how a story will work in editorial, the post-vis process creates animatics to fill in the sequences until they can be replaced with the real VFX composite.

Additionally, pre-vis can also lead to Tech-vis, designed to aid in visualizing how certain scenes may be captured, how sets need to be constructed, and what the final visual effects process might entail. A typical tech-vis delivery would include animated Quicktime movies, usually in plain view, encompassing the movement of the camera, lens, frustum, and the camera's visible path, along with the relevant dimensions and speeds. Ideally, these are tied together with the pre-vis shot, as a picture-in-picture, showing the pre-vis shot as approved, as well as the view through the tech-vis camera, which helps validate the tech-vis.

In pre-production, the VFX supervisor will determine whether blue-screen or green-screen will be used in the backgrounds on set, depending on what is planned for wardrobe and makeup for the actors. Also, wardrobe tests will be reviewed to determine whether issues may be found with fine detail in the clothing that might cause issues downstream. For digital CG characters, planning is required to figure out what should be shot in order to help the creation of the CG character in the VFX process. For the digital CG characters, reference materials are usually gathered so the VFX artists can see how real objects work (like animal movement).

VFX in Production

During production, tracking markers will be placed in the set to help the VFX vendors track the location and motion of actors and objects. Along with camera lens information and/or metadata, the tracking markers help VFX vendors reconstruct any motion to look like it was shot during production, with the same lens imperfections as what was used during shooting. VFX vendors will mimic the lens imperfections in their perfect CG renders in order to allow them to match what was originally shot.

Multiple witness cameras (video cameras) are used to shoot while the production camera(s) is (are) shooting. Witness cameras create three-dimensional data/depth to help with motion tracking of actors and objects which is used to help the VFX process in post with additional reference for more natural motion/perspective. After the set has been used for production (usually, later that day), the VFX team will capture High Dynamic Range Imaging or HDRI to help reference where the lighting sources were located during production. Facial scans will also be done during production to capture the actor's facial features and movements to help with creation of digital doubles and CG.

One thing to remember is that *any* changes from the original planning with the VFX team can drastically alter the budget – usually upward! Therefore, changes in the production need to be communicated with the VFX team in order to minimize budget changes for the VFX processing.

The VFX editor and editorial team track all of the elements that will be used on VFX shots and have the important job of pre-compositing or "pre-comping" various elements of a VFX shot together.

They might choose a certain background and layer/composite in other elements like people, objects or other elements to create an early concept of what the shot will look like for the VFX supervisor and other filmmakers to review. Once the pre-comp is approved, the elements that are needed for the composite are determined by the VFX editor, created and then sent to the VFX vendor in a process called a "turn over."

VFX in Post-Production

Moving forward to the post-production phase, the VFX editorial team will create the sequences and export EDL data. The EDL will contain metadata that references the original camera files, as well as the scene and take metadata. The company or department looking after the original camera files perform a "VFX Pull" – a process where original camera files are extracted and pulled together using the EDL data. This limits the VFX work to just the shots that are of interest to the VFX editorial team, and alternate shots and image sequences are not brought online, avoiding unnecessary or duplicated work.

The VFX house could work from the editorial files, but typically, they work off the highest resolution possible (like the camera files or uncompressed frames made from the camera files) if the workflow allows for it, as has been in the case of some recent theatrical and episodic productions.

Depending on the delivery specifications needed, the camera file is usually converted to a different file format in order to be used as a Plate to "turn over" to the VFX company (hence the term, VFX Turn Over where the plates are turned over to a VFX company so they can start their work). Camera RAW files or compressed camera files are also sometimes delivered as plates to the VFX company. This allows the VFX company to convert the image should their internal processing require a specific format or process in their pipeline.

However, if many different VFX companies do their own conversions using many different processes, there will almost certainly be discrepancies in the image quality between shots. Sharpness and dynamic range can be compromised making shots from different companies very difficult to match especially if the VFX shots have the same actors or easily identifiable landmarks. For this reason, it is important to establish clear guidelines on where the debayering will occur, and which tool (and version of the tool) will be used for debayering. In the case of some studios, baseline guidelines are provided for the production. If the content was shot on film, the film is scanned and files are created for the pulls.

The Plates are typically uncompressed 16-bit half float OpenEXR or 16-bit DPX frame sequences. Along with the plates, the VFX editor for the production will also provide corresponding information about the shots known as a count sheet which contains the length of the shot, camera report information, and other pertinent information from the VFX database about the shot as well as the pre-comp elements.

The VFX editor at the VFX vendor will receive the count sheet and pre-comp and will disseminate the information to various departments at the VFX vendor in order to get the work started. The VFX editor at the VFX vendor will compare the plates/scans with what was delivered in the pre-comp to check for errors and may create a pre-comp with the larger plates for the VFX artists to use as a starting point or for reference.

Sometimes the VFX supervisor will require the plates to have a neutral or balanced color grade before they are turned over to the VFX facility. More information about this can be found in the Digital Intermediation section under "Color Pre-timing and Reference Look." The process of generating and compositing VFX is complex, but more information can be found in the Special Section 3: Visual Effects.

Sound

Sound or Audio post-production is an elaborate process with several key groups involved. Each group is a microcosm of the whole post-production process under an editorial supervisor who works to the production's overall post-production supervisor.

The groups include dialogue, music, sound effects, foley, and mixing. In each sound group there are teams of specialists who work on specific areas. Although some of the sound work happens in parallel to the picture editing, much of the sound work happens once the edit of the content is "locked" where the timing and order of the shots does not change. The music team starts early by working with the composer and editorial to create the "sound mood" for the production. The editor(s) work to a combination of guide tracks and pre-mix sub masters as needed in order to ensure the cut and music complement each other.

Generally, waiting until picture lock prevents unnecessary effort on sound that might not be needed or has different requirements for the final edit.

Sound Editorial

The sound editorial Supervisor schedules the time to prepare the elements that make up what will be the final mix of the sound. Sound editorial usually breaks down into different element groups:

- Dialogue
- Music
- Sound effects
- Foley

Each group is responsible for creating their part of the sound. When previews or early test screenings are needed, each team will quickly prepare some basic elements for something called a "temp dub" which is then mixed within a few days (typically three to four days). The temp dub is a temporary mix that is used to convey what is intended for the soundtrack, but is not the final mix of the content, and the temp dub is made before the picture is "locked."

The next sections detail each of the groups that create the various elements.

Dialogue

The audio recorded during production can be used, but it often needs to be cleaned up to reduce or remove location background noise. The Production Dialogue Editor does this cleanup work using the cuts from the Picture Editor. If the production dialogue is too noisy or cannot be used, a process called Automated Dialogue Replacement (ADR) is used to replace the production dialogue.

Actors are brought into a special ADR recording studio to re-record their dialogue while watching a repeating loop of the image that will be used. The term "looping" is synonymous with ADR because the actor repeats the take until the new recording is synchronized to the lip movement and intent of the scene. The ADR Re-recorder does the recording of the dialogue while the ADR Editor edits the take to make it match even closer and to prepare it for the final mix.

Music

Like sound effects, the music can include pre-recorded songs or newly created music. A Music Supervisor oversees the pre-recorded music that is used in the edit of the content. This can include not only the main music in the soundtrack, but music that is in the background like on a radio or television. A Composer creates new music or score for the content which can usually set a certain mood or feeling during specific scenes. The Music Editor and Music Supervisor will go through a process called a "spotting session" where they will determine what sort of feeling or mood is needed during certain times of the edit. The spotting session information is given to the Composer so he or she knows what sort of incidental or transitional music is needed. Once the Composer creates and records the music, the Music Editor synchronizes all of the music to the edit of the content.

MUSIC AND EFFECTS TRACKS FOR TV

There is a lot of confusion around the meaning of Music & Effects (M&E) when applied to TV programs. Much of the confusion stems from the different interpretations and from the delivery requirement documents. M&E for TV is also genre dependent.

The first thing to establish is "what is actually needed." Don't assume just because one distributor or broadcaster has supplied a definition of M&E, that the definition can apply to all.

Two of the most common variants that come under the often-misused term M&E are:

1   Documentary M&E: The variation here is caused by programs with in-vision dialogue (i.e., interviews or presenter to camera) and programs with off-camera narration only.

    Where there is in-vision dialogue, some distributors require the in-vision dialogue to remain but out of vision to be clean. This gets really confusing where dialogue either starts or ends in-vision but sections are off-camera. Some want the dialogue cut immediately, others want it cut at a sentence pause, and there are others who want it kept. The "rules" can be really complex, so the best advice is to ask for a written explanation, and then save as many stems as possible during the final audio mix so it is possible to rebuild if needed!

2   Drama M&E: Drama M&E can mean separate music and sync/effects mixed, or separate music with separate effects and a dirty dialogue track which is dialogue with any location spill, or it can mean a total footsteps track with separate music, separate effects, and a separate clean sync dialogue track OR any combination of the above. The same rule applies as given on Documentary M&E: if you are unclear, ask for a written explanation, and then, save as many stems as possible during the final audio mix so it is possible to rebuild if needed!

Other than the above, the mixing of a TV program differs very little from a movie – except for the vast difference of scale and time!

Sound Effects

The picture editor will often use temporary sound effects to fill "holes" in the soundtrack. These effects need to be replaced by the real or specially created sound effects. The Sound Designer on the show creates new sounds that are needed for the content that might not appear in real life. For example, perhaps there is a new object that does not really exist – what should it sound like? Or perhaps a new alien world has been created from CG – there are no associated sounds with the world, so the sound designer would create the sounds to go with the image to elicit a certain "feel" for the environment. Not all sound effects have to be designed; some sounds can be found in pre-recorded sound effects libraries. A Sound Effects Editor takes all pre-recorded and newly recorded sound effects and syncs them with the edit of the content.

Foley

This process is named after Jack Foley.

Depending on how noisy or clean the production audio was, common everyday sounds like footsteps, clothes rustling, and door creaks may need to be re-recorded. These sounds are re-created and recorded by a Foley Artist. A Foley Editor takes the new sounds and synchronizes them to the edit of the content.

Mixing

Once all of the dialogue, effects, and music are recorded and synchronized to the edit of the content, the Sound Mixer can start to adjust how each of these sound types interact with each other including volume and placement in the sound field or fields required.

Predubs

Before starting the final mix, a "predub" is done to the dialogue and sound effects. Predubbing prepares the dialogue and sound effects for the mix by grouping them into similar groups in order to reduce the number of tracks to make it easier for the mixer to use and organize. Also, during the predub, any production audio is cleaned up and any fixes that are needed are made during this time to prepare them for the final mix. The audio levels of each track are evened out during the predub stage as well. Predubs usually take a few days to complete.

Final Mix

After predubs are finished, the mixer starts to adjust how each of these sound elements interact with each other including volume and placement in the sound field.

For theatrical content, mixes typically take four to six weeks but the budget and schedule determine how much time is allotted for the mix. Music is usually brought in at this point by the sound editorial music team and mixed together with the dialogue and sound effects predubs.

During the Final Mix, separate dialogue, music, and effects mixes are created, comprising what are called, "stems." The combining of the stems or composite becomes the final soundtrack of the content. Stems are kept separate, just in case fixes or adjustments are needed to each of the elements.

Depending on the type of combination, the sound field could be a simple mono or stereo mix, but the more common type is a surround sound mix like a 5.1, 7.1, or immersive mix. The sound mixer may have other mixers working with them to help run the multitude of tracks or elements of the sound. The mixer creates the final version of the soundtrack but will also create a soundtrack that only has music and effects (no dialogue) which can be used as a base for localized language tracks for different countries. Each mix is a different soundtrack. For theatrical content, most soundtracks are still done in reels as each reel is locked.

The director reviews the mix and notes any changes needed. The mixer will then work through the notes updating the mix. When everything is ready, a "sound check" is done so studio executives and production crew can review the soundtrack and determine whether any other fixes or creative changes need to be made.

When the final master mix is ready a last "print check" review happens in order to check for any technical issues. The term "print check" is a holdover from when film prints were checked for audio issues. The print check is really important because this version of the mix will be used to make all of the downstream masters – getting it wrong could lead to fixes being applied to many sub-master copies.

Soundtrack Versions

When the master version soundtrack is complete, all of the other versions can be created. Here is just an example of some of the versions that might be required.

Localized Dubbing

Localization is when a different language dialogue is created and mixed into each language's localized master. There is an art associated with proper localized dubbing where not only does the translation of the original language need to keep the original intent of the dialogue, but the actual words used should have (as close as possible) the same mouth shapes and movements of the on-screen actors. Making the translated dialogue/dub look like the actors are actually speaking the localized language, keeps the audience less distracted and able to focus on the story.

In a similar process to ADR, actors who speak the localized language record the translation which is then mixed into a localized soundtrack by the mixer.

Descriptive Audio

Descriptive audio, also known as "Audio Description" in Europe and "Descriptive Video Service" by US broadcasters, is meant for those who are visually impaired. The Descriptive Audio track is a separate audio channel that usually includes a narrator who describes any action happening on-screen in-between dialogue from the original soundtrack. Creating a good Descriptive Audio track requires a lot of preparation in order to make sure that the narration paints an accurate picture in words and during the gaps in the action dialogue.

Nearfield Mix

For theatrical content, often a nearfield mix is created for distribution of the content to the home. Home theater systems are different from theatrical systems, so a special home theater or nearfield mix can be created in order to optimize the soundtrack for the home. Typically, the adjustments are done to the surround channels for nearfield mixes since the levels differ between theatrical systems and home systems. These mixes sometimes reduce the level of the surround channels (or rear groups of channels) by 3 dB. But the real test is listening in a "home" environment during QC (see QC section).

Audio Loudness for TV and the Home

Broadcasters are continually getting complaints that commercials and trailers are too loud while the programs are too quiet, even when traditional meters show they are within the levels that the broadcaster has asked for.

There are many documents that try to explain the issue and suggest remedies but the thing to remember is, TV loudness is regulated in many countries, meaning there is no "negotiation." Programs either meet the local regulations or are "normalized" to meet them. Normalizing a finished mix will always be at best "OK" but usually is a disappointment and never meets the creative intent of the content creator.

To try and resolve the problem, new ways to measure audio levels have been developed. Two examples are:

- Much of Europe uses the European Broadcasting Union's EBU R128 recommendation. European broadcasters have unusually introduced this voluntarily which allows then to vary the requirements for special events; however, these variations are rare and should never be assumed until confirmed by the program's commissioners.
- In the United States, loudness control is mandated by the Commercial Advertisement Loudness Mitigation (CALM) Act, and non-compliance will fail QC and risks program rejection. The Standard used is ITU-R BS.1770.

The good news is there is very little difference between the two options which means, automated conversion between EBU R128 and ITU-R BS.1770 (in either direction) is usually acceptable, but there are always exceptions so never assume the conversion is OK until it's reviewed with the best audio device on the market: "Ears."

DIGITAL INTERMEDIATE

For content that is intended to finish in a quality higher than what was used in editorial, the final step in the post-production chain is the Digital Intermediate or DI process. The DI process encompasses digitizing the captured motion picture data, aggregating and conforming the original image files with visual effects and data from other facilities, such as title houses, and then color grading the images for exhibition. The DI process also includes delivery of the master files and data from which downstream deliverables will be generated, the manufacture of exhibition elements, and the creation of final archive data for the DI master elements. A DI post-production facility provides several key features for finishing the content including conform, image fixes, color correction, aspect ratio changes and final render for distribution masters.

Conforming

Conforming is a process where the images from the locked edit are matched back to the high-resolution version of the images (camera files and final VFX comps). In order to perform the conform, the DI facility uses the final EDL from editorial since the EDL provides a mapping back to the original camera files and VFX comps. During the conform process, any transitions such as fades, dissolves, flipped images, and others that were done in editorial are performed on the high-resolution version. Back when film was edited, these sorts of transitions were called, "Optical Effects," so even in today's digital world, these transitions are still called "Opticals," even though they are recreated in real time, digitally. The conform is the higher resolution version of the editorial media and becomes the base for the rest of the finishing.

Image Fixing

Now for the bad news! When the Conform is reviewed by the editor and director, they are seeing the images for the first time in their full glory... or not! Many times, once the film-makers see the higher resolution version, image artifacts become more noticeable. Most DI facilities have the capability to make various fixes to the image including painting out digital hits from cameras, vanity fixes (where skin of actors and actresses are made blemish-free) and grain/noise management for scenes that had too much noise either due to underexposure of the image, excess film grain, or other issues.

As any fixes at this time in the process can be very expensive, delay premieres and even worse, miss publicized delivery deadlines, the Post Supervisor needs to consider the tradeoff between delay, cost, and processing cost. For TV workflows, higher quality editorial files allow issues to be seen and addressed early in post but use more storage when there is more material being used. Lower quality editorial files need less storage and are faster to process but the risk can include missing image errors that could delay delivery while they are being fixed.

Finishing

When the Conform has been signed off at the DI facility, the job of "finishing" can start. Finishing can be a simple color match plus burned-in text and credits for a low-cost TV program, or it can be a wide range of high-cost processes, high cost as the images are now full resolution masters.

Color Timing/Grading for TV Programs

For TV programs color timing, sometimes called "color correction" and in Europe often referred to as "grading," is a stage between the Conform and the final edit to add graphics and text before export and localization edits.

The three most important parts of the color timing process are

1   The colorist
2   The display
3   The room

The EBU technical document TECH 3320 outlines the requirements for displays used for TV production, and the ITU has possibly the most famous document on testing and viewing environments, ITU-R BT.500 "*Methodology for the subjective assessment of the quality of television pictures.*"

The display and the room setups are critical when color work is made. The initial color correction is done for a specific display type. For episodic content, the color correction is usually targeted for home television displays. There are three primary options for consumer TVs:

1 SDR HD TVs – these only reproduce ITU-R BT.709 images
2 SDR UHD TVs – these should additionally reproduce ITU-R BT.2020 images
3 HDR UHD TVs – these should additionally reproduce ITU-R BT.2100 images

Only "broadcast quality" Grade 1 TVs should be used for color correction. Grade 1 TVs or "monitors" can fully reproduce the images from the standard they are fed. This means that artifacts and errors are not masked by processing in the monitor and that they can be used as a measurement for the visual evaluation of image quality. One of the important factors is that Grade 1 monitors have good color stability (the colors will not drift during the day) and should reproduce images accurately based on the image standard being viewed.

The only exception might be a color matching process where one or two shots are being matched to others. Usually, this occurs when the original material is fully graded or has come from a Studio or outside broadcast truck when all cameras have been matched and new material is introduced during the edit that does not match. In these cases, the color-timed material can be used as a reference, and the new material is matched to this reference.

A full color session for TV programs usually takes between three to five times the program duration, assuming relatively well-matched cameras were used during the shoot. However, if the content comes from unmatched sources, archive, other programs, it can take up to ten times the program duration to achieve the creative intent of the production team.

If the TV program has only one delivery format, usually standard dynamic range (SDR) HD, then a single color session is all that is needed. If a high dynamic range (HDR) or an OTT platform (Netflix, iTunes, etc.) is needed, then you need to plan for two or more "passes."

One thing to note: in most of Europe, if the master high dynamic range version is Hybrid Log-Gamma (HLG), then a single HDR pass is all that is needed. The backwards-compatible image capability of HLG allows a UHD SDR TV to display a reasonable representation while a simple 3D LUT can be used as part of a down-conversion to generate an SDR HD version. This is usually the workflow for live TV programs but can also be used for non-live content.

Color Timing/Grading for Motion Pictures

Color correction (or color timing or color grading) is where the bulk of the time and effort occurs for the DI facility. The conformed image is loaded into a color correction system and is controlled by a colorist, a person that helps translate the Director's (and sometimes DP's) vision for the content into the actual image. Color correction can be as simple as balancing a scene between two actors shot at different times to make it look like they were shot at the same time to changing the look of the shot completely. Colorists can change the color to elicit certain feelings during a particular scene. For example, if the filmmaker wants the scene to feel like a warm summer day, the colorist can add a bit of red or orange to "warm" up the image. If the filmmaker wants the content to feel cold and icy, the colorist can put a blue cast on the image to make it "feel cold." If the scene was shot during the daytime, the colorist can change the image to make it look like the action is happening at night. Another example of what colorists do is when a production shot in color is turned into black and white by the colorist, but certain objects are left in color. The colorist has the ability to change the color on certain objects or people in order to create a certain look as requested by the filmmaker. This change is permitted through the use of "power windows" which allow an object or person to be isolated from the rest of the image and have color only affect what is within that window.

As a side note, the above process can also take place on the editorial files if a production is not required to deliver the content in any format higher than the editorial format, for example, for preview purposes. The color correction can use the dailies grade as the starting point or can start from scratch to get a completely different look than the dailies grade, depending on what the filmmaker desires.

In comparison to TV episodic content, theatrical content is a bit more complex because the initial color correction or "Hero Grade" is targeted for a theatrical projector (in DCI P3) and additional trim passes are done for other devices like home television displays. Both types of content can have additional complexity with the advent of high dynamic range (HDR) displays both in theatrical and home environments which usually requires additional "trim passes" for each different display technology. A Trim Pass is a separate color correction pass that is done to optimize the image for a specific display device and is based upon the hero grade.

For example, the hero grade would have the daytime scene turned into night optimized for a theatrical standard dynamic range (SDR) projector. The theatrical environment is very dark, with a large screen, but the projector only outputs a maximum of 48 cd/m^2, so this hero grade would get a simpler grading pass or trim pass done to it to optimize it for the home television. The home environment is typically a brighter environment, with a smaller screen that has a much brighter maximum of 100 cd/m^2, in the case of Standard Dynamic Range, and could be either 1,000 cd/m^2 or 4,000 cd/m^2 in the case of High Dynamic Range. The image is usually modified slightly to look good in this different environment and display. Otherwise, if you used the theatrical hero grade on a home display it would look dark and dull. Trim passes are also done to make the image look better and convey the proper artistic intent for other display formats including HDR theatrical, 3D stereo theatrical, HDR home, and 3D stereo home.

No one way of finishing a movie exists. For many theatrical features, below is a typical order of how the grades are completed:

Theatrical

1   2D Hero Grade (SDR)
2   3D SDR at 3.5fL / 12 cd/m^2
3   3D SDR at 7fL / 24 cd/m^2
4   3D SDR at 14fL / 48 cd/m^2
5   2D Dolby Vision Grade (HDR)
6   3D Dolby Vision Grade (HDR)

Home

1   2D Rec.709 Grade (SDR)
2   1.78 Pan-Scan (SDR)

3 3D Rec.709 Grade (SDR)

4 2D HDR Grade

In parallel, IMAX, Eclair or other special theatrical distribution companies would receive files to create their 2D and 3D grades for their deliverables. Note that some productions start with the HDR theatrical (Dolby Vision) as their hero grade and then create the other trim passes. For some content, the hero grade for the home masters is an HDR home (Dolby Vision) grade that becomes the starting point for the other trim passes.

One of the most important tasks during the digital intermediate process is to maintain the color correction through editorial changes. The first set of digital intermediates were carried out on "cut-negative" shows. For these productions the editorial cut had been made and locked, essentially "cutting the final negative" and preparing it for the answer print stage. In this case, while the colorists had to make several creative choices, they were not hampered by shots being replaced/changed while they were setting the creative look. However, today the editorial process is often in process and rarely locked when the post-production schedule mandates that the digital intermediate process begin. It then becomes necessary for the DI colorist or color assist to work with a DI editor to ensure that color correction can proceed on a previously locked cut, and when a new version of the locked cut is released from Picture Editorial, the new shots are introduced into the timeline without losing the color correction that was already carried out on previous shots.

A new editorial cut could insert, move, shift, delete, or even shorten or lengthen existing shots in the time new shots are replaced into the timeline. It could also replace a shot in its entirety, e.g., in the case of an upgraded visual effect shot that replaces an earlier temporary placeholder. This can be a fairly cumbersome task, as simple color corrections such as global corrections can be easily copied over and shifted based on the changes in frame count. However, in the case where "power windows" and special tracking of objects is employed, it may be necessary to manually go into each new shot and adjust the tracking and power windows.

Color Pre-Timing and Reference Look

As mentioned in the Dailies section above, the DIT and DP will take the established reference looks and may do some simple (and occasionally complex) color correction to change the look of shots or to balance them within a scene. Part of the DI colorist's job is to help establish the reference looks in a color-calibrated environment with the DP. These reference looks are often referred to as the color bible for the show, and from this bible, the DP has a set of looks for the common situations of daylight exterior, daylight interior, nighttime exterior, nighttime interior, and any special looks that he/she might be interested in maintaining. Part of the objective is to make the dailies process as smooth and streamlined as possible. Additionally, it also helps to convince the DIT and DP that small global modifications using the ASC CDL are sufficient to balance out shots. Without this confidence, the DIT and DP may employ higher orders of color correction, like power windows and secondaries which would then require the DI colorist to spend a lot of time undoing these corrections and balancing things out, to ensure a consistent pipeline across a production.

Another important aspect of the DI colorist's work is to pre-time the visual effects plates as they are delivered to the VFX companies. Given the large percentage of VFX in current motion pictures, it is important that shots delivered to each VFX company are balanced, so that once composited with computer graphics imagery they can easily match one another. The DI Colorist/Color Assistant is responsible for balancing and or pre-timing each of the shots that are "pulled" for VFX work, so that there is a uniform look to each one. While this is an extra expense and extra step, it ensures that all shots are in close proximity to the final look, and when returned to the DI they will match closely.

Mastering and Versioning Preparation

The hero grade and the trim passes each have to be rendered out from the color correction system before being used as a master for distribution, but before that is done, Aspect Ratio conversions may be necessary for the Home Entertainment master.

The aspect ratio is determined by the width of the image divided by the height of the image. Theatrical content is usually shown in either scope (2.39) or flat (1.85) aspect ratio, while most home televisions have an aspect ratio of 16:9 (1.78). There is also a plethora of handheld devices with varying aspect ratios and many 4:3 (1.33) displays still exist.

For the home master, an Original Aspect Ratio (OAR) version is generated, i.e., a version that preserves the theatrical aspect ratio, and also a version that plays back properly on a 1.78 display.

There are actually two masters formatted for 1.78 displays. For the first generation, the entire OAR image is fitted horizontally with black bars at the top and bottom of the image. The second fills the screen horizontally and vertically which some viewers prefer. This means a decision must be made as to which section of the OAR is shown. The theatrical content goes through a process called "pan and scan" or "pan-scan" in order to create a 1.78 full frame image. The DI facility is equipped with the ability to basically take a 1.78 window and pan it around the 2.39 image in order to create a 1.78 master. Inevitably, image content will be cropped out, so changes are sometimes made to the cut in order to properly display an actor who is speaking.

Once the pan-scan and color passes are approved, the DI facility can render out the color and pan-scan versions according to the delivery specifications given by the content owner. The hero grade, the trim passes, and any aspect ratio changes are separate renders. These renders become the image portion of the masters for the various distributions like theatrical or home. The DI facility will also render out files for archive.

Figure 4.8 shows a simplified workflow for the DI where the source material, regardless of whether it was shot on film or digital, is brought into a color correction system. The original ASC CDLs from dailies could be added if desired, but the key is to show how the hero grade is created using Step 1 which targets a certain type of display output, in this case, the SDR theatrical projector. The grade is typically done through a look and output transform that limits the grade to the capabilities of the display device, again, in this case for a P3 theatrical projector. Once the hero grade is complete, it is rendered out, in this case, for a Digital Cinema Distribution Master or DCDM. Now, Step 2, the HDR theatrical projector grade, can be graded through a trim pass based upon the hero grade. This is accomplished by changing the output transform to something different that allows the grade to meet the capabilities of the HDR projector. Then, the color for the HDR projector grade is rendered out. The same process is repeated for Step 3, the home Rec.709 grade, and Step 4, the HDR home grade, and for each new version, the trim pass is done off of the hero grade. Once the project is complete, an archive is rendered out from the hero grade, but note that typically

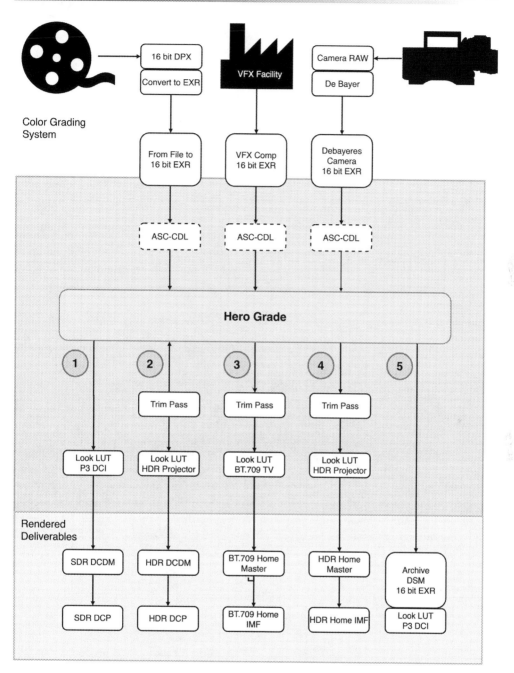

FIGURE 4.8 This Diagram Highlights the DI Post-Production Image Workflow for Feature Post-Production

the rendered files do not include any looks or output transforms since those would limit the range of the image to the display device that was used for the hero grade.

Mastering and Versioning

The DI facility or a mastering facility will take the image renders from DI and add any audio, captioning, and subtitling to create the final masters for distribution. Depending on the type of show, this is where the versioning nightmare begins! For theatrical workflows, masters are created for Digital Cinema and then for the home distributions. For Digital Cinema, the DI facility will render out either a Digital Cinema Distribution Master (DCDM) or compress directly to a Digital Cinema Package (DCP). If a DCDM is created first, a DCP is created from the DCDM. The DCP is a standardized format that is accepted by all Digital Cinema theaters (although there are countries that use a different format) and is essentially a lossy JPEG2000 compressed version of the movie, with all of the audio channels held within an MXF wrapper. The maximum bit rate of the JPEG2000 images is constrained to 250 Mb/s. More details on the DCP construction can be found in the Modern Digital Cinema Workflow section below.

The DCP format allows for flexible versioning which allows for the image and audio to be ordered and played back by a playlist called a Composition Playlist or CPL. The CPL references or "points" to the image, audio, and other files like captioning or subtitles which allows many versions to be created without having to have multiple instances of the image, audio, or other media files. More details on the CPL construction and uses can be found in the Modern Digital Cinema Workflow section below.

As described earlier in the workflow, many versions exist for Digital Cinema since there are many distributions. Several Hollywood Studios have reported that they have released more than 350 different versions of the same movie due to localized versions, different picture versions (2D, 3D at different light levels, Dolby Vision, IMAX, etc.), different audio configurations (5.1, 7.1, Theatrical ATMOS, DTS-X, Barco Auro, etc.), and different edits due to censorship in different countries. But this is just for theatrical! The home/non-theatrical space has easily just as many different masters for a theatrical title due to similar situations with localized versions, different picture versions (2D, 3D, HDR), different audio configurations (2.0, 5.1, 7.1, Home ATMOS (which is different than a Theatrical ATMOS mix), DTS-X, etc.) and different edits to follow the censorship cuts in different countries. In addition, different resolutions (4K UHD, HD 1080, HD 720, etc.) increase the number of versions that need to be created.

For Home Entertainment and other downstream channels, rather than a DCP, an Interoperable Master Format (IMF) package or some other type of file-based master is created. As of 2018, many theatrical productions are created in uncompressed formats or IMF for archive or higher-level servicing while Apple ProRes QuickTime files or IMFs are created as smaller servicing files for distribution servicing.

Interoperable Master Format (IMF)

The Interoperable Master Format (IMF) is an internationally recognized SMPTE standard for the interchange of file-based, multi-version, finished audio-visual content. The IMF was loosely based upon the DCP standard but was meant to solve the versioning issue for home masters. IMF has the same basic structure as DCPs where a CPL references image, audio, and other files, allowing multiple versions to be created through a playlist. In addition to versioning, IMF supports the complex home ecosystem with different resolutions and frame rates within its architecture. Although IMF standards began to publish in 2012, as with any new standard, adoption took a while, and started gaining popularity in 2017.

IMF is not a format for day-to-day program production and post-production. The non-linear editing systems used during program production are storytelling tools that treat each video and audio element as discrete items whose relationship is set by the storyteller's intent. IMF was originally designed to allow high-quality content to be exchanged between businesses. It requires tools that manipulate and understand finished components of a program and skills that are very different to those used in traditional post-production.

IMF is a workflow format that allows, or rather, enables all the additional elements of a TV program or movie to be "bundled" together in an open and flexible way.

A typical non-linear edit application (NLE) could make any version of a program from the master file, but they are not specifically designed to do so. IMF tools are ideally designed to organize finished content segments, or, as the IMF documentation describes them as "components" and compile the different elements into a finished version. More detail about how IMF can be used in post-production can be found on the SMPTE website (www.smpte. org).

Mastering and Versioning Applied to TV Documentary

There are many similarities in the mastering and finishing processes for movies and TV programs destined for an international market. It is worth looking at how processes are becoming more streamlined and use a number of automated processes.

Figure 4.9 gives an overview of a typical "high end" documentary – shot on a variety of cameras each using the best codec for the particular camera. Shooting ratio is typically 800:1 to 1,000:1 and all rushes are kept in their native format, for use by the production company of other programs or for potential "stock-shot" sale to other program makers. There are no particular "preferred" processes but the use of mezzanine compression during post-production is kept to a minimum, or if not possible, a camera "native" end-to-end codec might be used. A mix of common tools is normal and devices like FilmLight Baselight and SGO Mistika are common to cinema and TV post. However, some processes such as HDR to SDR conversion, HLG to PQ conversion, etc. are often totally automated and occur during the creation of the final format masters.

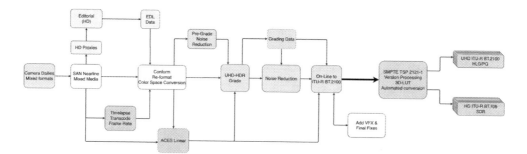

FIGURE 4.9 Television Post Workflow

TITLING, CAPTIONS, SUBTITLES, AND ALTERNATE TAKES

Title Sequences

The opening title sequence and end credits are typically created by a title facility. The studio or content producer usually has a legal team that vets all names and credits of people that worked on a particular production. These title sequences have to be created similar to VFX content and delivered to editorial and the finishing post-production facility. The length of each of these title sequences is determined by the content producer. Depending on the production, the title of the content may be localized in different countries and created in the same style as the original title.

Some broadcasters only allow "approved" job descriptions on programs they have commissioned (and paid for). Always check the broadcaster or distributor credit guidelines before final post.

Captioning and Subtitling

Once the edit is locked, translators can finalize the caption and subtitles for the content. Captioning is visual words on-screen that is usually meant for the hearing impaired and includes the dialogue, descriptions of sounds, music and who is speaking if the person is off screen. Subtitling can either include the same content as captioning or it can be a translation into another language. Subtitling is often a different translation from localized audio in the same language because the subtitle language does not have to account for mouth movements and can be a direct translation of the dialogue. Subtitling does have to take into account how many words can fit on the screen and still be human readable, so there are constraints that a subtitling facility uses when creating the subtitle. Both captioning and subtitling are usually delivered as certain types of file formats that are usable for downstream distribution.

Alternate Takes

The first question you should ask is "what are alternate takes?" and the second should be "what has to happen to the alternate takes I have been asked to process?"

Alternate Takes are quite simple technically. They are just different versions of a shot or clip of sound (or both) that replace a similar shot or clip in the program. The simplest form would be a shot with burned-in text (e.g., a contributor name) and the alternate take would be a clean version with no burn-in. But an Alternate Take can also be a unique location shot or VFX sequence. Also, there may be more than one! An example of this is in animation, where alternate versions of a sequence where background detail or text is changed for either language or cultural reasons.

When planning the Alternate Take options, it must be clear what exactly is meant. For example, do ALL of the shots with text need to be clean and to what level? This could mean graphics such as maps or story-line critical background text (as opposed to background text that just happens to be in shot) or it could be as simple as clean versions of that shots that can be used to replace "dirty" versions.

The same applies to sound; however, this may mean a new mix (see the Audio Mix section).

QUALITY CONTROL (QC)

QC for Theatrical Delivery

Quality control for theatrical content must be done in a properly calibrated theater. The first QC is done off of what should be the final Original Version (OV) DCP. The OV contains one editorial version of the content and is usually the base for localized languages. However, there can be multiple versions of the content (for example, the English OV may have on-screen text burned in while an international OV may use textless image, expecting that D-Cinema subtitles will overlay over the textless image). Also, keep in mind that each version can have multiple audio and subtitle languages, as well as multiple audio configurations such as 7.1, 5.1, Dolby ATMOS, DTS-X, etc. Each of these variations must be reviewed for quality control! Starting with the OV DCP, each of the variations (audio configurations, captioning, language versions, etc.) are QC'd. Then, any other OVs are reviewed along with each of those variations.

The typical order for the review of the DCPs is as follows:

1   QC OV DCPs (all combinations of image & audio)
2   Apply any necessary fixes or updates, and QC updated DCPs
3   Once a particular OV is approved, the package is used as the source for versioning all DCPs for this particular picture & sound combination...and this process starts all over again when there is a new DCP version!

Checks are performed on image and sound at the same time. For 2D content, the review is usually done in a theater with a white screen. For 3D standard dynamic range content, the review is often done on a silver screen. Although the facility making the DCP will also QC the content to make sure there are no technical issues, the Studio will often have several people that are familiar with the content review the DCP to make sure there are no other issues, creative or technical with the sound or picture. Usually, this includes at least one crew member from Sound (mixer or sound editor) who is familiar with the mix and at least one crew member from Picture or DI (Director, DP, colorist, Editor, Assistant Editor, VFX Editor, etc.) familiar with the cut and one familiar with the color grade, as well as someone from the Studio who is overseeing the particular title.

QC is often the last opportunity to make fixes or changes before mass duplication starts and dozens (sometimes hundreds) of versions are made from the OV. Therefore, this is the time where filmmakers and crew members squeeze in last minute updates. While the intent of the QC process is to correct things that are "wrong," many times it is also utilized as the last chance to make changes.

For some filmmakers the QC period is a good time to review the content in a new environment. For some people, they finally get to view the content on a much larger scale while others who are unfamiliar with the content are asked specifically to review the content with "fresh eyes." These new factors sometimes offer a better opportunity to identify previously unseen or unheard issues. For example, an editor may have only seen their content on a 50-inch monitor in editorial, but once they see the 2K or 4K DCP projected onto a 50-foot screen, flaws are much easier to spot.

QC is done to identify, and hopefully fix any technical, production, or creative issues. Picture issues can include compression artifacts, duplicate frames, dropped frames, frame edges in picture, conform issues (wrong shot or VFX version), power window issues, bad or old color, 3D phasing, incorrect or misspelled credits, soft or fuzzy on-screen text, production issues (boom/production equipment or crew in frame), poor speed changes, incorrect placement or timing of on-screen text and subtitle or captioning issues.

Sound issues can include sync issues, misaligned reel breaks (resulting in a pop or snap), pops or crackles, incorrect channel routing, missing audio, incorrect sound mix balance issues, Narrative Audio Description issues (for visually impaired DCPs) and hearing impaired DCP issues (boosting of the center/dialogue channel). For dubbed versions, checks would include making sure the correct language is playing back in addition to the above issues. The different audio configurations must be reviewed on the correct type of system. For example, a 7.1 mix must be reviewed in a theater with the proper system to decode and playback 7.1.

Prior to the visual QCs, a data validation of the DCP is done to check the interoperability or proper playback across various devices, as well as confirming that the DCP has been mastered correctly. Additionally, the studio can request that the mastering facility review the DCP and confirm or reject based on its own technical specifications and requirements.

QC for Television

QC for most broadcasters usually consists of two distinct operations:

1   Compliance QC (CQC)
2   Content QC (QC)

Compliance QC for Television Delivery

A "compliance check" is to confirm the file is the correct codec and format. The compliance check, sometimes called "compliance QC" (CQC), is totally automated and checks the file formatting only – in other words, "will it play and is the metadata formatted correctly?"

Broadcasters usually have a delivery Portal or Gateway for content delivery and the CQC checks occur in the gateway. Files that are OK are passed to the next step while files that fail are "bounced" along with a technical report detailing the failure issues. In the UK the Digital Production Partnership (DPP) has a common CQC for all broadcasters. The details are published, and many automated QC vendors include the "UK DPP AS-11" CQC test. This allows post companies to pre-check file before sending.

Content QC

Assuming the program passes the CQC, it moves to the broadcaster's QC department. Again, this is split into two processes:

1   Automated QC – AQC
2   Human QC

AQC carries out the checks that humans cannot or where humans use devices like video waveform monitors or audio level meters. Humans are expensive and QC areas even more expensive! Many QC checks that require meters and signal level monitors can be automated which means humans only need to QC sound and picture content and do not need to watch a program with technical errors.

AQC devices check for technical compliance with the video and audio standards the broadcaster has asked for and also the format of the file matches the requirements. Automated QC (AQC) devices usually work to QC Templates that are either supplied by the broadcaster or meet the house style of the post-production company. Many broadcasters and distributors are supporting the EBU QC site to define their QC Items and QC Templates.

Human QC is a visual and audio check by experienced craft personnel capable of assessing picture and sound quality, in an area where there is no distraction, controllable lighting, and reasonable soundproofing, on the final flattened transmission file. It is not acceptable to view any layered version of a program.

The viewing is carried out using a good quality broadcast video monitor Grade 2 or better with a minimum size of 24 inches, plus correctly positioned good-quality speakers.

A Grade 2 is similar to a Grade 1 but usually has wider (or looser) specification tolerances than a Grade 1. Grade 2 monitors are a cheaper than Grade 1 and are used where the tighter tolerances of a Grade 1 monitor (for example on accuracy of color reproduction and stability) or the additional features of a Grade 1 monitor are not needed.

Some facilities use "high end" consumer televisions fed via HDMI. There is something to be said for watching a program on the best version of the target display! If the following guidance is treated as a set of "minimum" requirements, then it is an acceptable procedure:

1   Minimum screen size 42"

2   Resolution*

    a.   HD – 1920×1080 native resolution with 1:1 pixel mapping
    b.   UHD – 4320×2160 native resolution with 1:1 pixel mapping

3   Connections

    a.   HDMI 1.4 or later for HD
    b.   HDMI 2.0b or later for UHD

4   Viewing distance, no more than 3H for HD and 1.5H for UHD with the viewing position directly in front of the screen preferably with the viewer's eyes perpendicular to the display horizontally and vertically

5   External high-quality speakers properly positioned for stereo or surround sound monitoring – see Figure 4.10 for 5.1 but Report ITU-R BS.2159 gives suggestions and diagrams for optimal listening for other formats**

6   Reasonable sound dampening to minimize external interference and reflections

7   Regularly calibrated with broadcast monitor test signals (e.g., PLUGE, Testcard, etc.)

8   Controlled lighting to minimize screen reflection

    *Resolution – it is always best to use a TV that is native to the target format, but higher resolution TVs can be used (3840×2160 to QC 1920×1080) IF and only IF engineering tests show no visible degradation due to the TV's internal scaling.

    **At the time of writing, one of the most popular Next Generation Audio (NGA) systems is Dolby Atmos – this adds height to the sound arena but only adds four additional speakers to a 5.1 setup as shown in Figure 4.10.

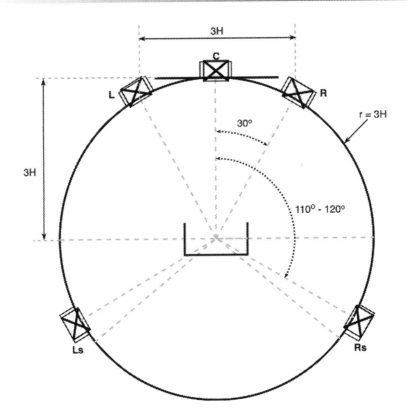

FIGURE 4.10 Speaker Configuration for Audio QC

Area for TV Immersive Audio QC

It is becoming increasingly common for TV to use more immersive audio formats such as DTS X and Dolby Atmos. When it comes to QC for these formats, there is no point in using a cinema style listening environment. The TV QC area has also to assume the audio and video will be QC'd together making the distance from the screen just as important as the position of the speakers.

There are two common types of room layouts:

- **Equidistant**, where the distance to each speaker is approximately equal. This layout works well when a surround sound QC area (as per Figure 4.10) is adapted for immersive audio QC.
- **Orthogonal**, where the room is usually longer than wide, more common for sound mixing where the mixer is toward the back half of the room. This layout can be used when the sound is QC'd in a sound mixing area.

Figure 4.11a gives the approximate speaker positions for an immersive audio QC area (including video QC). The height of the top speakers is usually about 2.5m from floor level and angled toward the operator's position.

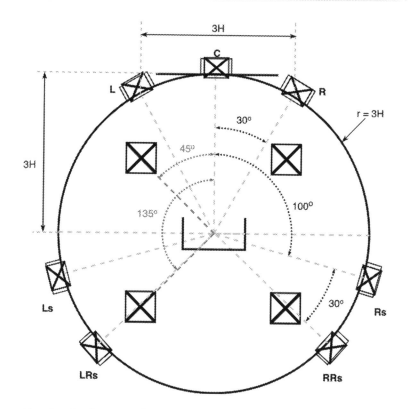

FIGURE 4.11 Speaker Configuration for Immersive Audio QC

QC Database of QC Items

The EBU has an open-source database for TV QC (AQC and CQC) which the UK's DPP and many other broadcasters use to define their QC requirements. (http://ebu.io/qc)

PHOTOSENSITIVE EPILEPSY (PSE)

Some countries require all programs to have a PSE test and report. This tests for flashing images and high detailed patterning. Details of the test can be found in ITU-R BT.1702. Many AQC devices have an option to carry out PSE tests but not all are approved for use in every country. In the United Kingdom, PSE tests are mandatory and approved devices are listed on the UK's Digital Production Partnership website

SOUND AUDIBILITY FOR TELEVISION

The quality of sound is an extremely import facet of the program, and poor quality can adversely affect the viewers' audio-visual experience. Poor sound is usually blamed on:

- Noisy location recordings
- Mumbling actors

- Too loud background music or effects
- Inconsistent dialogue levels between and within programs
- Quality of television speakers

Many TV viewers complain they have to adjust the volume during a program: this is often referred to as "volume surfing," i.e. the scenario where viewers turn up the volume to be able to understand what's being said but then have to turn the volume down when a loud section of the mix happens, usually music or fast action. Because the volume has been increased to understand the dialogue, the action sequences become too loud!

Audio mixers utilize devices that can help to ensure the final mix in within the local regulation or guidelines (Europe uses EBU R128 and the USA uses ITU-BT.1770). There are other variations used by other regulators or broadcasters, so it is very important to get the details as early as possible.

For the final mix, the music is normally taken down 4dB, and this has been shown to make an enormous difference to the audience and can usually be achieved without affecting the creative intention.

Additionally, a regular recommendation is to be aware of is how television sound is transmitted. Usually only the surround mix is sent to the home which means the TV or set-top box will do the down mix for those with stereo (usually TV speakers). If the mix doesn't take account of this, the dialogue, which is fine in 5.1, can get swamped in stereo.

Some tools that might help assess a mix are available. In order of preference:

1 Trained Ears!
2 Simple Loudness Meters
3 Sound analysis tools (for file and live programs)

There are many Loudness tools now available as plug-ins to editing software as well as dedicated audio mixing tools.

As an example – take a look at the analysis evidence of two programs which both meet the European EBU R128 requirements on loudness, but the first example has good dialogue reproduction and the second caused Volume Surfing (Figures 4.12 and 4.13):

ARCHIVE

With all the creative input from the production, post-production, and creative teams, the final deliverable becomes an extremely valuable product. It is extremely important to archive it in a fashion that permits easy access, long-term reliability, and data protection. Additionally, based on specific studio guidelines, one or all of the below are archived:

a. The Digital Source Master (DSM) is archived. This is the final version of the graded files from the DI process;
b. The Ungraded DI files, the DI project, and all Viewing LUTs are archived;
c. The Original Camera Negative (or RAW Camera Files) and production audio are archived;

Loudness graph against time

Histogram Display

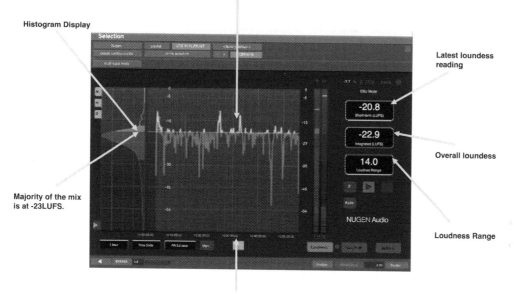

Latest loundess reading

Overall loundess

Majority of the mix is at -23LUFS.

Loudness Range

NUGEN Audio

Timescale in minutes

FIGURE 4.12 Example Showing Good Dialogue Reproduction

Loud sections that will cause Volume Surfing

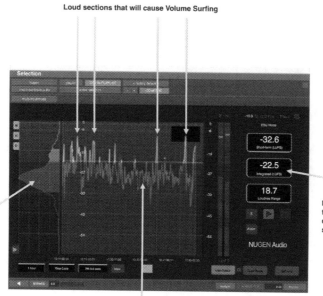

Mix is within EBU R118 tolerance but far too quite due to very loud sectiotions

Majority of the mix is below -23LUFS.

NUGEN Audio

Too much of the mix is below the EBU R128 -23LUFS level which means the dialogue will sound too quiet

FIGURE 4.13 Example Showing Material That Leads to Volume Surfing

d. The Visual Effects Models and Textures are archived;
e. The Print Master as well as the Audio stems are archived;
f. The Digital Cinema Package (DCP) is archived: This is the final "release" print of the motion picture; and
g. The Interoperable Master Package (IMP – the file that represents the IMF) for the Home Master is archived.

The DSM represents the final creative intent, and in the case of multiple theatrical versions, each of the light levels as well as the SDR and HDR version are archived. The ungraded DI files, the DI project and the viewing LUTS represent all the files before creative adjustments were made for a specific display medium. Given the rapid advances in display and audio technology, this data set can be repurposed for the next generation audio-visual experience. The original camera negative and production audio were captured and backed up on-set and these are preserved in the archive, as are the Audio Print Master as well as the Audio stems. In certain cases, the DCDM from the HDR grade is also archived, as are the models, textures, and references generated during the VFX process.

Typically, these assets are received on LTO data tapes, and the archive sets each have a well-defined naming convention, along with Metadata requirements, and checksums for each TAR (tape archive) group to ensure that data validation schemes can be run across these data tapes. These data tapes are written in either the TAR or LTFS format and image files are normally written with the following guidelines:

- Image files must be written to tape in sequential order as they appear in the film.
- Sequences of files may be broken down into reels or sequences, as long as the correct sequencing information is maintained in the metadata and filename.
- Image sequences are to be contiguous when written across multiple tar volumes.
- Each file, sequence and/or reel on a given tape must be complete, with the following guidelines:
  - No frame file shall carry over to the next tape in the set,
  - No sequence or reel shall carry over to the next tape in the set,
  - If a full sequence cannot fit on one data tape, the sequence may span to the next tape in the tape set as long as any given tar volume does not span the tape set.

- For a 3D title, left and right eye of the same reel may reside on the same data tape sets as long as the reel does not span across another data tape. If both left and right eye cannot fit on same data tape, they shall then be written to separate data tape per reel.
- Use No Data File Compression.

  - Hardware, software, or proprietary data compression may not be applied when writing archival data tapes in order to ensure native drive capacity.

- Approximately 100MB of the tape must be left data-free to guard against tape handling errors.
- The end of each tape must contain an "end-of-data" file mark indicating the end of the recorded data.
- A file mark must be written after each tar volume indicating the end of the "tarball."

THE FUTURE OF POST-PRODUCTION

Across the entire media industry there is general move from large capital infrastructure operations to "on-demand" operational cost-based processes. As connectivity speeds rise so does the viability of cloud and diverse location collaborative post-production. This is already happening with audio post and is possible during editorial, where proxy files are small enough for real-time operations.

Currently, most of post-production processes (dailies encodes, editorial, etc.) are carried out in post companies data centers; this is known as on-premises or "on-prem" operations. This is because the high-quality image files are very large and speed is the imperative factor. However, some areas of post operations are able to exploit cloud-based and off-site operations. Rendering for VFX is starting to move into the cloud because of the spiky nature of the work – we all know sometimes there is a massive amount of rendering to be carried out and then sometimes there is very little. On-prem facilities would need to have peak capability ready to use 24/7, which translates into high investment in hardware that may have very low utilization.

Cloud services offer the ability to spin up more processing when needed and turn it off when not needed. Instead of investing in on-prem infrastructure that might sit dormant for most of the time in the anticipation that it will be required to support peak demand, paying for the cloud processing when you need it can be an efficient and cost-effective workflow.

Other parts of post-production are starting to look at the cloud. Transcoding for dailies and VFX plate pulls seem to make sense since you can spin up processing when needed, but the drawback is speed – for example when processing dailies, moving large camera RAW files into the cloud in the first place especially from remote locations is not viable because of access speeds: Network connectivity to the cloud from various locations can be limited or prohibitively expensive. Most productions do not have the budget for a super-fast connection to the cloud which means that large files like camera RAW files can take a long time to move into the cloud. Because productions need dailies files turned around quickly, waiting for your camera RAW files to move into the cloud before dailies processing is time-prohibitive right now.

Yet even now in News production, cameras are connecting to cloud services and uploading content as its shot. Many cameras can already produce viewing proxies simultaneously with the "masters" and this might well be the first step in long form production in the cloud. Mobile 4G can be used but 5G networks can be locally set up to allow local to cloud transfers during shooting time. Cloud processing could mean the program's look can be preset, then accessed and used by all companies which will eliminate the differences that occur when multiple vendor processing occurs.

Once your source data is in the cloud, the possibilities for manipulating the files are endless. Many companies are creating Software as a Service (SaaS) that can be used on a pay as you go service to process files. Machine learning services can interrogate the images and audio and provide additional metadata such as image and sound recognition to your files and there are various transcoding and image processing services available.

The potential for better remote collaboration is a key tenant of working in the cloud. Many people are familiar with working on a document in the cloud with others such as a Google Doc or Microsoft 365 Word document, and these SaaS offerings are being joined by others such as DropBox, Box, etc.

These companies are looking into how very large files can be processed without the need for transfer, production-based backup, and built-in version control. It is also worth considering that the Cloud operators can manage vast processing power in an on-demand way, and many already offer real-time proxy streaming based on the recipients' network speed. It would not be science fiction to suggest that soon an IMF CPL could be "dropped" into the production's cloud space and a real-time proxy of the version stream back to a Director's cell phone or a full quality version made available to a buyer with no need to transfer.

As has been demonstrated by the Audio community, especially in music composition and recording, there are amazing opportunities for collaboration with many people working on a project at the same time in real time. Now imagine that you have a set of image and/or audio files in the cloud and everyone can be editing or viewing the files in real-time at the same or different times. You would have true remote collaboration for the storytelling and for reviewing and approving the content.

Currently there is a problem that needs to be addressed, lag or as it's known, "latency." Vision switchers, editors, colorists, etc. expect – well demand – that when they press a button or adjust a setting, the change occurs on the next frame (mid frame changes are a bad idea!). For Cloud production to become viable for the end-to-end process, this latency issue must be resolved. It must be remembered that while Cloud processing can be much faster and much cheaper than on-prem, control signals and "return" suffer latency.

Even with current high bandwidth connections, the latency or delay in receiving the data over a network connection makes it difficult to work in real-time. With the image files in the cloud, the latency from the network connection would cause a delay in seeing the change. For cloud operations to become fully viable for all areas of post-production, post areas will need to invest in much higher connectivity for both local and cloud connection.

One other item to be aware of with production in the cloud is that companies that provide traditionally on-prem processes are aligning themselves with one of the multiple cloud providers. At some point, many of these services could become cloud agnostic, but because each cloud provider has different technical infrastructure, it makes sense for companies to work and hone their product or service to optimize for a specific cloud provider's framework. For example, Avid® is currently working with Microsoft Azure for the implementation of their Media Composer editorial software in the cloud. This means that productions working in a different cloud might have to support multi-cloud usage and may have to move files from one cloud to another. As of the writing of this book, a new business model for the cloud providers will probably be needed in order to make production in the cloud cost-effective and a reality for content creators.

One other worthy note to include in the future of post-production is the use of game engines to provide real-time rendering for VFX, as opposed to the non-real time rendering that is used currently (aka start the render and get a cup of coffee/entire meal). Game engine companies have already demonstrated VFX quality rendering on short pieces, and their quality and usability will only improve in the future.

BONUS CONTENT

Special Section 1: Color Pipeline

Each production needs to determine how color will be processed from camera through to the finishing process. Many productions have a few choices here including using a workflow specified by one of the camera manufacturers, working in a specific color space like Rec.709 or a standardized workflow like ACES (Academy Color Encoding System). It is important that the color workflows/pipelines be determined before production starts in order to prevent problems with color mismatches downstream. Testing the color workflow from camera through various areas like dailies/editorial, VFX, and DI using camera test footage before production starts helps to validate and test the color workflow and determine whether changes need to be made.

The color workflow/pipeline through finished content includes the following items:

- Debayering of camera images into a format to be used as the source for the post-production process (if using camera RAW files)
- Color used for dailies/editorial
- ASC CDL values that adjust the image – can be per shot or sequence
- Overall look of the content (more of a "film look" or a "higher contrast" look)
- Viewing pipeline for VFX and DI (and which display device will be used for the hero grade)
- Whether or not there is a neutral grade for the VFX work
- Hero color grading working space

The above items show a sample of where color can be modified in the workflow of post-production. However, color can be altered whenever a file is converted into another file or if the images are manipulated while viewing them on monitors that have not been calibrated to the proper standard. Since maintaining creative intent is such an important part of creating content, paying attention to the color pipeline is a requirement. Otherwise, you may hear complaints like, "That's not what I shot!" or "That's not my movie! That snow isn't the right shade of white!"

Proprietary Color Workflows

Each camera manufacturer has their own proprietary color workflow that can be used by a production for their color pipeline. For example, if you shoot on ARRI cameras, a production can use the ARRI LogC workflow which converts the camera into an optimized color gamut that ARRI has specified for their camera images. Besides providing a special color working space, ARRI also provides transforms that can be used for different display devices, so if your hero grade is targeting a Rec.709 television, for example, you could use their transforms to view the color on a Rec.709 monitor. Once you establish a look for dailies, you would want to make sure that you communicate any looks and transforms as well as the target display device in any deliveries to VFX facilities to make sure they are viewing the images properly for dailies color. If you are targeting a Rec.709 finish, then the color management can be quite straight forward with the final DI/color correction. If the final color correction is targeting a different display device, for example, a Digital Cinema projector, you would need to use an ARRI transform for a Digital Cinema projector as your viewing transform for the hero grade.

Many dailies, color correction, and VFX systems understand the LogC workflow and can work with LogC images. However, Canon has their CanonLog workflow, Panasonic has their VLog workflow, Red has their RedLog workflow, and Sony has their SLog3 workflow. If you shoot with several types of cameras, it is possible to pick one of these camera workflows for all of the cameras, even though it sounds counter-intuitive to do so. Many filmmakers insist on having a certain look to their images and choose to work in one of these proprietary camera manufacturer–based color spaces. It should be noted that not all of these proprietary color spaces have been documented, so although they will work in the short term, there is no guarantee that these color spaces can be recreated in the future. From an archive standpoint where you may pull out your Digital Source Master 50 years from now, you would want to make sure that you can re-create any approved color or looks. One of the difficulties we face is if the proprietary camera transforms or working spaces are not available, you will not be able to get an exact match.

Besides the camera-specific workflow, there is also the issue of the look being proprietary to either the creator that made it or the post-production facility that made it. Back in the film days, certain post-production facilities prided themselves on the various ways that they would output film. Creatives would go to one post house over another because of the way that post house's film-outs looked. Once Digital Intermediates became popular, post houses needed a way to emulate how these digital images were going to look on their film-outs, so they created Film Print Emulation (FPE) looks. The trouble was that these FPEs were quite proprietary to the post-production facility, and the post facility would not include this FPE with the DSM archive. This meant that when content owners restored their DSM, they could not get the exact approved look because the FPE was not supplied, and without it the images would not look right. You could also render in the FPE into the DSM, but then your image would be limited to what the FPE would allow. Although most productions do not output to film as their hero grade anymore, the post-production houses were used to not including any looks with the archive files, so many digital only productions still have this problem today. The post-production houses are now starting to include the looks/transforms, so always make sure to ask for them so that the looks can be added (virtually) in order to create the approved hero grade.

ACES

The Academy Color Encoding System version 1.0 was published in 2015 by the Academy of Motion Pictures Arts and Sciences (AMPAS) as a framework that can be used to place any camera image into a standardized container and color space. ACES is meant to help with color management from set through finishing and into the archive. Because it has been standardized through the Society of Motion Picture and Television Engineers (SMPTE), ACES provides a standard "language" to define color and looks that were used on a particular production. Where proprietary color spaces are not the best for archive purposes, ACES strives to be a standard way of defining color and looks for image archive. The ACES standards can be found in the SMPTE document suite of ST2065.

Through a series of transforms called input transforms, the camera files are converted into the ACES files, also known as ACES AP0 EXR files. Each camera manufacturer is responsible for making the input transforms according to the guidelines set by AMPAS. The ACES files can in turn have other transforms act upon them to change the look or optimize the output for different display devices. A look, for example, a more contrast-y look or a film look, can be stored in a transform called the Look Modification Transform or LMT.

There are different output transforms for each display device. The basic concept is to use the output transform that matches your hero grade, for example, the Theatrical standard dynamic version. Once the hero grade is complete, you can swap out the output transforms for a different display device, for example, a Dolby Vision projector, the Rec.709 HD display device, or an HDR television monitor, which will optimize the grade for that particular display device and can be used as a starting point for your trim passes for each new display device. The ACES framework allows for the color transformations to be broken up into components that can be added "virtually" over the ACES image so that the image never has any looks "baked" or rendered into the image. This concept of not baking anything into the image is important since many looks (including LMTs) have a tendency to limit the dynamic range of the image, locking it in to the display device's capabilities at the time it was approved. Display devices are always improving, so if you were to "lock" in a specific look today, you would not be able to get more out of the image (Figure 4.14).

ACES includes a few "working spaces" for specific usage in color correction systems and at VFX facilities. In color correction systems, there is a working space called ACEScct that allows colorists to have their controls on their color correction system to "feel" like they normally do. For VFX facilities, ACEScg provides a working space that is optimized for use with computer graphics. ACEScct and ACEScg are not meant to be rendered out and exchanged; instead they are meant to be only working spaces that are virtually created. A DI post-production facility and VFX facility would render out ACES AP0 files to exchange or use for archive.

There are two other components of the ACES framework that help with interchange and basic human usability. ACESclip is a sidecar metadata file that is used to describe the transforms used and the order that they appear. This file is helpful for archive and can serve as a record of what was done to the ACES files in order to re-create the approved look of the content. Looks are stored in a file format called a Look-Up Table or LUT, but because there was no way to interchange them between systems, ACES created the Common LUT Format (or CLF).

ACES also defines an entire workflow using film where film can be scanned into files that can be transformed into ACES and can also be recorded out to film from the ACES files. As of the writing of this book, the latest version of ACES is version 1.1, although many systems still use version 1.0.3 which does not include the newest output transforms for HDR. The ACES system is a work in progress and a current effort called ACESnext is working on improving the features and usability of ACES. AMPAS and SMPTE have defined a standardized Digital Source Master archive file through the use of an IMF (Interoperable Master Format) application that uses ACES image but includes synced audio and subtitles and other metadata. This format can be found as SMPTE ST 2067-50.

Special Section 2: High Dynamic Range

High Dynamic Range (HDR) refers to the contrast range of the displayed image. The most common mistake people make is that HDR is all about brighter images. HDR is more about

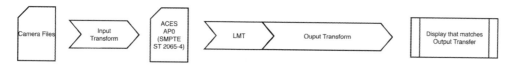

FIGURE 4.14 Basic ACES Workflow

darker images, where it enables the display of more detail and color in darker images areas plus incredibly bright highlights.

It should be mentioned that, at the time of writing, HDR is a nascent technology and experience is still growing. Any description of HDR in the workflow is a snapshot in time and the reader should consult regularly updated reports and standards from organizations such as the ITU and SMPTE.

In Theatrical exhibition, the Standard Dynamic Range (SDR) is normally 2000:1, or about ten stops. For SDR television in the home, with a nominal peak at 100 cd/m^2, the dynamic range is about 128:1 or about seven stops.

In Theatrical Exhibition, there are now various projection systems such as IMAX, Éclair, and Dolby Cinema that offer different contrast ranges for the displayed image, and as they are greater than the standard Digital Cinema specification, they are regarded as High Dynamic Range systems.

For consumer televisions, there is a wide range of display technologies, such as organic light-emitting diode (OLED), Liquid Crystal Display (LCD backlit with Light emitting diode (LED)), and so on, and each of these display technologies have different capabilities and thus different dynamic range.

The good news is that the ITU Recommendation ITU-R BT.2100 defines only two HDR options for program production, Hybrid Log-Gamma (HLG) and Perceptual Quantization (PQ), and standardized under SMPTE standard ST-2084. This may seem confusing especially when documents refer to ITU-R BT.2020 color as well, so it's worth a brief overview.

ITU-R BT.2100 has "cut and paste" color text (literally) from ITU-R BT.2020. This means:

- if a UHD program is HDR, ITU-R BT.2100 is all that's needed to describe Color and Dynamic Range; and
- If a UHD program is SDR, ITU-R BT.2020 is all that's needed to describe Color and Dynamic Range.

There is significant confusion in the industry currently over the different types of HDR, including what is actually meant by HDR. This should not affect post-production unless a proprietary distribution HDR format is used too early in the workflow.

Transfer Functions

In post-production, the basis of HDR is established by the Transfer Function. Recommendation ITU-R BT.2100 standardizes two HDR transfer functions: HLG and PQ (SMPTE ST 2084). The report ITU-R BT.2390 gives more background detail to both systems, and ITU-R BT.2408 gives operational guidelines. Additionally, SMPTE ST 2084 defines the optical transfer function for mastering reference display for the PQ HDR system, which is not needed for HLG.

Theatrical workflows tend to favor PQ workflows while TV favors the HLG workflow. The main difference between the two is that PQ is a "display referred" and provides a specific code-value to luminance correspondence, while HLG is "scene referred" and the images have the same relative and the actual display brightness is not relevant.

As long as everything has been captured in either a camera RAW format or RGB (4:4:4) then HDR post-production is the same as Standard Dynamic Range (SDR). Wherever possible, master HDR content using ACES or RGB 4:4:4; if this is not feasible, choose the mastering HDR format and stay with that format throughout post.

The good news is, ACES, RGB 4:4:4, HLG, and PQ do not need any form of metadata during post.

HDR Signals

It is important to keep in mind that since a larger range of the image will be displayed on an HDR display or projector, the bit depth of the camera image should be large enough to provide the desired dynamic range. For theatrical workflows 14-bit cameras are required, with the image bit depth based at a minimum of 16 bits. For TV broadcast workflows, the minimum is 10-bit but 12-bit is preferred if heavy post-production processing and color correction is needed. It is pointless to process HDR in 8-bit; it is simply not acceptable!

For Theatrical finishing, the HDR master is generated while viewing the material on an HDR projector. There are instances where a 42" Dolby Pulsar has been used as a starting point to generate the trims, but predominantly the colorist will begin the HDR grading process directly on the HDR projector for which the session has been allocated.

To preserve the look of the motion picture, typically the show's look is transformed into the larger container of the HDR display, so that the colorist has a starting point to take advantage of the deeper blacks and brighter highlights.

For the Home Entertainment master, the colorist will normally color grade on an HDR display. Two common displays used in the mastering process are the 42" Dolby Pulsar and the Sony BVM X300. Based on production guidelines, the colorist will select the peak luminance point, and if using the Pulsar, this is set at 4,000 cd/m^2, and if using the X300 it is 1,000 cd/m^2. This does not mean that the content will be forced to either 4,000 or 1,000 cd/m^2; it simply means the "container" permits this. In reality, the creative choices will determine how bright or dark the content actually will be.

HDR Displays

Probably the most important piece of the HDR process is the display used for color grading or final post. Monitoring HDR is not as straightforward as getting an HDR capable display and setting it to the right version! First the display needs to be set up and the room needs to have the right amount of illumination or the pictures will not look as intended when they are distributed.

A new display line up PLUGE signal has been developed by the ITU and a revised version of ITU-R BT.814 has an explanation of how to use the signal to set up the display brightness.

What is important to remember is the way a display should be set up. It is important to follow the guidelines in ITU-R BT.814 and adjust the controls in the correct order and in an area with lighting conditions.

When making graphics and adding text to HDR programs a new "white" level reference must be used. Adding text at 100% white level will make it far too bright with respect to the pictures so-called "Graphic White" must be used. Report ITU-R BT.2408 has tables with details of skin tones and graphics levels.

Brightness Jumps

Cuts between shots with very different brightness can be as uncomfortable as sudden transitions between quiet and very loud audio. Large brightness changes from dark to light can be used sparingly for effect but cutting from a very bright sequence to a very dark action will not work as expected! It is well known that the human eye takes time to adapt to the dark so if there is action or detail in a dark scene after a hard cut from a very bright scene no one will be able to see it. Adaption can take up to 30 minutes in extreme cases – just try reading print after coming in from bright sunshine to a dimly lit room! It is important to ensure that the brightness variations within HDR programs are limited and controlled to avoid viewer discomfort.

The ITU operational guidelines suggest, "night scenes will usually have an overall brightness at the lower end of the normal operating range, and sunny outdoor scenes will have an overall brightness at the upper end of the range." It also points out that low light detail may be lost after a transition from a bright scene even if the transition is not uncomfortable, because it takes time for the eyes to adapt.

Line-up

Finally, a new television format will need a new set of color bars... ITU-R BT.2111 defines color bars for PQ Full and PQ Narrow Range and HLG Narrow.

The new signals have traditional HD colors as well as HDR colors.

This is an example of the new HLG color bars (Figure 4.15).

Conversion for Television Programs

Many television programs, especially documentaries use content from a wide variety of sources not under the control of the program maker or simply from archive content suppliers. The ITU Report ITU-BT.2408 "Operational practices in HDR television production" details some of the processes and practices HDR operations.

FIGURE 4.15 Example of New ITU-R HDR Color Bars

Cases to consider are:

1. Converting HDR content to an SDR signal range (i.e., Reducing the dynamic range of content).

   a. Display-referred tone-mapping is used when the goal is to preserve the artistic intent of the original HDR ITU-R BT.2100 content when shown on an SDR ITU-R BT.709 or ITU-R BT.2020 display (UHD with no HDR). An example of which is the conversion of HDR graded content for distribution on an SDR service.

   b. Scene-referred tone-mapping will change the appearance of the HDR content after conversion, but it is useful in live TV production where HDR cameras have been shaded using their SDR outputs. Scene-referred tone-mapping will produce a signal that is very similar to the SDR signal used for shading, and one that can be inter-mixed with other SDR cameras.

2. Placing SDR content in an HDR signal without changing its dynamic range.

   a. Display-referred mapping is used when the goal is to preserve the original "look" seen on an SDR ITU-R BT.709 or ITU-R BT.2020 display, when the content is shown on an ITU-R BT.2100 HDR display. An example of which is the inclusion of SDR graphics in an HDR program, where color branding of the graphics must be maintained.

   b. Scene-referred mapping will change the displayed "look" of content, but it is useful in live production where the goal is to match the colors and tones of a BT.2100 HDR camera. An example of which is the conversion of SDR graphics that are required to match in-vision signage.

3. Increasing the dynamic range of content by placing SDR content in an HDR signal with expanded luminance range, thereby providing a better match to native HDR content.

   a. Display-referred inverse tone mapping is used when the goal is to preserve the artistic intent of SDR BT.709 or BT.2020 content, when the content is shown on a BT.2100 HDR display. An example of which is the conversion of graded SDR content for distribution on an HDR service.

   b. Scene-referred inverse tone mapping will change the displayed "look" of content, but it is useful in live production where the goal is to match the colors and tones of a BT.2100 HDR camera. An example of which is the inclusion of specialist SDR cameras (e.g., slo-mo, spider cams, robo-cams) in a live HDR production.

4. Convert SDR content to HDR and then back to SDR or HDR content to SDR and then back to HDR. While such ability is very useful, it is advised to limit the number of such repeated conversions as much as possible, as some signal degradation is likely.

Many ask:

* Why would you convert SDR to HDR then back to SDR? and
* What has this got to do with post-production?

If you consider the end-to-end workflow of a program, it becomes clearer (Figure 4.16):

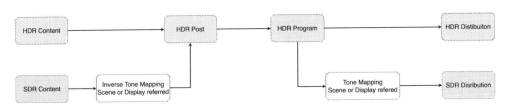

FIGURE 4.16 HDR-SDR Round Tripping

To ensure that image quality is not unduly compromised due to image processing, each step in the end-to-end chain should be mindful of processing that has already occurred or will occur later. The old engineer's mantra "*it's ok leaving me*" is not good enough if there is no consideration to the next step image processing.

Special Section 3: Visual Effects

Asset Creation

The VFX company has several steps in order to create their final approved VFX shots. The basic process for creation of something that did not exist before is to:

1   Create the object, character, or background
2   If needed, animate the character, or apply motion capture of the actors
3   Add textures to the object and look development
4   Set up the environment and light the scene
5   Develop effects like water, fire, smoke, dust, etc.
6   Composite the various elements together
7   Render the composite into a final form

Creation of the object, character, or background is dependent on what is needed. For example, a background could be a matte painting which is a painting of a background that needs to be composited into an existing shot. Because far away backgrounds typically do not need to move with foreground objects or actors, they can be 2D or flat in nature. However, in order to create foreground objects or characters that move around in a scene, steps must be taken to create the object or character so it can be seen from all sides and move around in a believable fashion.

In the case of 3D models and characters, it is necessary to develop the model/character from scratch, including geometry.

Model

To create a digital CG object or character, the first thing to create in the computer is a model using 3D software. The modeler sculpts the model as if it were made of clay, but the methodology and approach to creating the computer model differs significantly in the sense that models are usually made up by combining many polygon shapes to create the model. Models are mathematical representations of the object or character.

Rigging

If a model needs to move, it must be rigged. Rigging creates a skeletal structure or "bones" for the model. Rigging is a pretty technical process and requires knowledge of programming to create the proper rigs.

Texture

The surface of the model requires a realistic texture which is created by texture artists in 2D first. The 2D texture is then applied to the model in 3D graphics applications by mapping the 2D texture onto the 3D model to give the model more realism.

Animation and Performance Capture

Once a model is rigged, and a skeleton structure is developed, the model may be animated. For nuanced characters, the limbs, eyes, mouth, clothes, etc. of the model are moved by the animator. Another approach is to use performance capture systems to capture the motions of a real-life actor in the physical world and transfer them onto the digital model, wherein the joint structure of the real-life actor corresponds to the skeletal structure defined in the rig.

Motion capture has become a widely used technique because of its efficiency as well as high fidelity in being able to capture the nuances of specific actors. In feature film, often the performance capture is fine-tuned by animators to bring out specific nuances of the digital character.

Look Dev

Look development is where the photo-realism of the model happens. During look dev, the lighting of the object and shaders are balanced out to match what is in the actual scene. The lighting artists work to mimic any on-set lighting by placing CG light sources in the same locations and angles as what was used during production. Placing the lighting in the same location as what was used in the scene allows the CG asset to integrate into the original plate. Shaders are the instructions that describe how a surface should behave when light hits it and includes what color or "shade" it should be, whether it is reflective or has a matte finish and can even alter the texture somewhat by creating bumps and other surface textures.

Compositing

Compositing is where all of the various elements of the image are put together. Many times, compositors have the tough job of making sure everything looks right. They may clean up some of the rigging or fix blue screen lighting contamination. They help by doing technical checks and quality control of the image before delivery and play a large role in making sure color is balanced throughout the image.

Rendering

Once the 3D model is created and ready to be incorporated into the scene, the model must be rendered out in 2D to be added to the original plate. Rendering can be done in passes where separate parts are output to allow for better control in the scene.

World Building

World building refers to building an entire location in CG. Typically, this environment would seem to cover a large area from far away, but as you zoom in and get closer, there is enough detail through the location that you can "walk through" various areas of the location. You can think of world building as generating an entire town, for example where you can walk around the buildings and even possibly into each building.

Simulations / Particle Systems

Simulations are 3D computer models that are used to generate certain types of realistic looking effects that might randomly change with forces like wind or gravity. Many simulation models use particles to create the effect where each particle has its own properties like a certain mass, color, speed of travel, etc. Examples would be simulating how cloth moves in the wind, movement of water or buildings crumbling after an earthquake.

As the shots are being created at the VFX facility, they are continually reviewed, first by the VFX facility and then by the filmmakers from the production. The filmmakers will send back any notes for the VFX facility to work on. Upon approval, the VFX shots are delivered to Editorial (as editorial media), so that they may be cut into the current timeline and viewed in continuity.

As shots are finalized, the VFX facility sends the higher resolution file to the VFX team for final review and to the Digital Intermediate post-production facility for use in finishing. For VFX shots that do not use plates from the actual production, as in a fully CG shot, the process of review and approval is the same, and the VFX house would deliver fully rendered frames in the resolution needed for finishing as well as files for editorial to use.

CHAPTER 5

Managing Your Media Assets and Workflows

Section Editor: Jay Veloso Batista

The next workflow steps media assets follow can be convoluted and specialized but increase the monetization of the material at every step. To assist your understanding, this chapter focuses on the criticality of unique identifications for your content, which logically extends to the fundamentally important area of metadata management, and business management implications, as monetization of assets requires juggling multiple media formats to maximize their applications. This chapter explores new forms of essence, such as Virtual Reality, Augmented Reality, and haptics. And to fully understand the workflows involved in the media chain, we conclude with a look at the various business systems involved, including Quality Control and repurposing of content. To increase the financial viability and profitability of the media assets, today's business systems focus on automating workflows, and this chapter is devoted to explaining the applications of the many diverse systems involved.

There is a well-worn advertising description that has become so widespread and applicable in the media industry that it now commands market research as its own category by the IABM (The International Trade Association for the Broadcast & Media Industry) – that catch phrase is "Best of Breed." Media production companies demand the top, best performing systems, and this market indicator can be measured year to year. Product vendor companies use this portrayal to indicate their product is the best in its category positioned versus competitors. As with many marketing axioms, the phrase contains a germ of truth – the truly best in class products are very complicated and leverage unique design attributes that allow them to be integrated into a complete system that performs better than the overarching all-in-one tools sold by some vendors.

In order to provide you a deep understanding of these various paths media assets can take, we have assembled a panel of expert contributors, each with industry credentials or leading product expertise. Each section highlights the expert contributor and, as much as possible employs their words to describe the impact and importance of their topic to the track of the assets through the modern workflows in today's media industry.

CONTENT IDENTIFICATION AND ITS OWN METADATA

Advertising

*When we discuss making money with our assets, a fundamental topic is advertising. Identifying advertisements to ensure proper use in broadcasts has been an important issue since the very beginnings of broadcast Radio and Television, and today those requirements include streaming media, Over-The-Top services, and Video-On-Demand playback. An advertiser only wants to pay for the proper advertisement in the proper program location. **Harold S Geller**, the Executive Director of Ad-ID LLC, a leader in the advertising industry and an expert regarding the history, systems and methodology, explains the basis and applications of metadata to advertising assets:*

"If you can't identify it, you can't operationalize or measure it; if you can't measure it you can't monetize it," Clyde Smith, SVP Advanced Technology, FOX has said. Unique identifiers and metadata for advertisements must be universally recognizable and employed across the industry to see the full benefits, in Operational systems, in Administration and Finance, and of course in Measurement. There are two registration authorities that supply the industry with advertising identifier metadata, and they are Ad-ID in the United States and Clearcast in the United Kingdom: Both provide unique identifiers, metadata, and APIs for validation and retrieval. By using IDs and data, and the associated program integration tools, media companies can maximize interoperability.

The benefits and cost savings of interoperability are most apparent when technologies work together so that the data they exchange prove most useful at the other end of the transaction.

We need highly interoperable systems at the technology, data, human, and institutional layers. See Figure 5.1.

In the United States, Advertising Digital Identification, LLC (Ad-ID) is a joint venture of the American Association of Advertising Agencies (4As) and the Association of National Advertisers (ANA). Ad-ID was launched in 2002 and replaced the ISCI (Industry Standard Coding Identification), which was in place since 1969, established by the Broadcast Administration Technical Developments committee of the 4As. ISCI was withdrawn from the market in 2007. As of 2014, SAG-AFTRA requires that all ads using union talent be coded with Ad-ID.

All Ad-ID codes are 11 characters except for HD or 3D codes, which have an H or D in the twelfth character. See Figure 5.2.

Ad-ID provides the Complete External Access (CEA) API to approved companies (generally media outlets, online publishers, measurement companies, and their industry vendors) with access to all Ad-ID codes and selected metadata in the Ad-ID system. Companies using CEA must know the Ad-ID code they are looking for in order to validate the code or read the associated metadata.

Ad-ID advocates and supports embedding the Ad-ID code into files, using file native metadata format, and subsequently using the CEA API to retrieve the necessary metadata. In file-based workflows, the objective is to automate as much of the process of moving digital files through the supply chain by integrating different business systems. Having the identifier embedded in the file, and having that identifier survive the various transcoding processes is of critical importance, as ads are used on multiple platforms, and multiple consumption devices.

Industry Benefits of Ad-ID include:

- Guarantee of Unique Codes: This foundation is vital for the digital era and needed by the entire advertising ecosystem, including advertisers, agencies, vendors, and media;

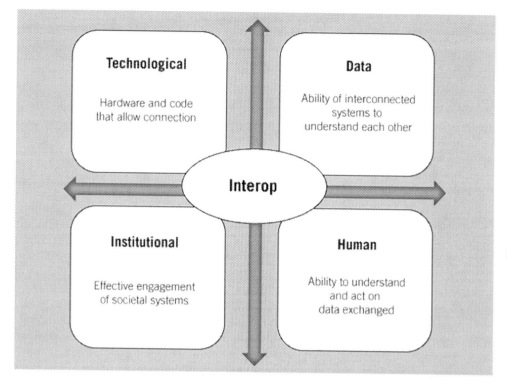

FIGURE 5.1 Interoperability Layers

What is Advertising Identification & Metadata?

Core Fields		Product Categorization	Code Information
Media Type	Length/Size	Industry Group	Ad-ID Code
Parent	Agency Name	Major Category	GUID
Advertiser	Language	Sub Category	Date Created
Brand	Copyright	Product Category	
Product	Version		
Ad Title	Commercial Delivery	**Media Specific Fields**	
		Video Format Flag	
		Bleed (Print)	
		Color Type (Print)	
		Expandable (Internet Display/Mobile)	

Company Prefix   User-Provided (letters, numbers)   High-Def or 3D

ABCD XXXXXXX H or D

FIGURE 5.2 What Is Advertising Identification & Metadata?

- Improved Reporting Accuracy: The Ad-ID system allows for enhanced evaluation of advertising assets;
- Removal of Re-keying and Duplication of Efforts: Without Ad-ID, an asset's identifying information must be re-entered into systems of record up to 20 times from when an advertiser gives approval to create an ad to the time the ad is actually published and invoiced. This is currently a huge duplication of effort, resulting in human error and increased costs; and
- Streamline of Workflow via Integration across vendor products: When advertising systems work together, the whole industry benefits. Ad-ID is integrated with a growing list of industry vendors driving advertising interoperability and simplifying processes throughout the industry.

We have learned from history – the fact is that non-standard, home-grown IDs for advertisements are bad for the Industry. These kinds of IDs do not account for the operational steps that follow the point where the ad leaves the advertiser's domain. Unregistered identifier codes cause confusion among the rest of the advertising, broadcasting, and media community, and lead to wasted time and money. When everyone has their own methods to track, execute, and measure, it takes everyone involved increased time and effort to move things forward. See Figure 5.3.

As the advertising requirements in our media channels grow more complicated, the Media and Entertainment industry needs greater transparency and standardization. The industry has to be able to efficiently identify, manage, and track creative assets. Using the Ad-ID metadata and enforcing its persistence in your monetization workflows provides a reliable

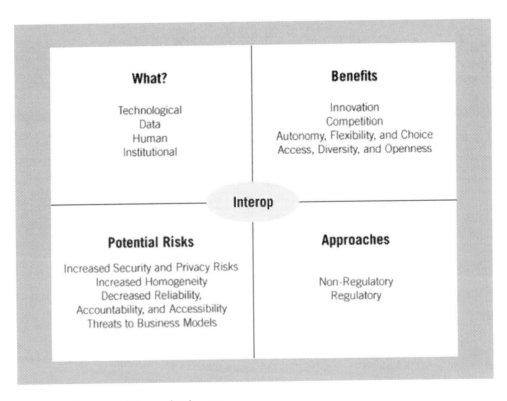

FIGURE 5.3 Interoperability and Advertising

business platform, providing better digital video asset data, coding, and distribution in an industry that is becoming extremely fragmented.

Content Identification for Programs

EIDR

The entertainment industry has always had its own unique media identification difficulties. Different companies and software management systems introduced wide variations in categorizing assets, metadata formats, and even version confusion, such as how to define the difference between an original theatrical release and a "director's cut," or how to manage the different language translations. A committee of industry executives founded a non-profit organization, the Entertainment Identification Registry (EIDR), to provide a shared universal identifier system for theatrical and broadcast television assets. According to their own promotional description, "From top level titles, edits, and DVDs, to encodings, clips and mashups, EIDR provides global unique identifiers for the entire range of audiovisual object types that are relevant to entertainment commerce. The Registry's flexible data model supports complex relationships that exist between various assets and is interoperable with other standard and proprietary identifier schemes."

EIDR was designed to supplement and support cross reference the application of existing asset identification systems for business to business media distribution needs as well as the requirements of the post-production community. The centralized registry is based on the industry standard Digital Object Identifier (DOI) technology and is available to all participants. It is easy to register media and create and ID and the IDs are "immutable" in respect to rights, or location of metadata or physical asset. The system was architected to prevent duplicates, both to support episodic series, and the ability to create complete sets of video assets from an original abstracted work, such as all the versions of a theatrical release.

The registry is built on a collection of records sub-divided into fields, with each record referenced externally by a DOI and once established every identifier is immutable. Underneath the DOIs is a Handle System, and each native EIDR ID is handle formatted, so the structure is increasing in its specificity, to handle, then DOI then EIDR standards. The resulting "IDs" follow a particular format:

$$10.5240/XXXX\text{-}XXXX\text{-}XXXX\text{-}XXXX\text{-}XXXX\text{-}C$$

where 10.5240 is the DOI prefix: The "10" indicates the handle is a DOI and the "5240" is assigned to the EIDR Association. The remaining string after the "/" is the DOI suffix where each "X" denotes a hexadecimal digit (A-F), and "C" is an ISO check digit. The EIDR registry can also supply a 96-bit Compact Binary form of the ID that is intended for embedding in payloads such as watermarks and a form for "web use."

EIDR supports four kinds of content records each with its own reserved prefix: "Content" for Movie or Television series assets; "Party IDs" to identify partners, media producers, and distribution companies; "Video Service IDs" to identify channels or networks; and "User IDs" to address the need to manage companies that are subordinate to the parties yet inherit the media access rights. Content records are objects categorized by their types and relationships with the expectation that the Content Record will be permanent, but the organization did plan for human errors and corrupted records – there is a mechanism for linked records called "aliasing" and a manner to designate a record a "tombstone" or dead record.

Another key aspect of the EIDR identifiers is that they support a rich set of Alternate IDs. This allows EIDR IDs to be integrated everywhere in content workflows. Alternate IDs are stored in the metadata of the EIDR ID and support the inclusion of proprietary as well as other standard (e.g., ISAN) references, and these additional identifiers can be added by any party that needs a new ID to support a new workflow. More details on the specification and the ID formats can be found at *www.eidr.org*.

Metadata

We mention metadata often, it seems encompassing and ubiquitous, so, what is this metadata? For our purposes of moving digital media assets through multiple workflows, our media must be annotated with additional data that enhances its applications and interoperability. For an in-depth analysis of this important aspect of managing our media workflows, we turn to **Bruce Devlin**, long time industry expert and the Founder/CEO of Mr. MXF and SMPTE's Standards Vice President – Bruce starts with a definition:

Metadata: *plural noun(**computing**): information that is held as a description of stored data.*

This seemingly innocuous definition hides an immense amount of complexity that has arrived over recent years in the definition, processing, distribution, and transformation of descriptive data. To really understand metadata (or data about data) in the timeline media world, it's worth looking back in time and seeing how we got to where we are today. This will help explain some of the mismatches between theory and practice that become evident as we discuss this topic.

Back in the days when physical media was interchanged, the metadata was generally helpful information that was easily interpreted by humans. Sure, there were databases of scripts and cast lists and media formats, but essentially all workflows were governed by the tape cassette or the spool of film. Metadata might travel with the physical media in the form of paper or a label, or maybe an ID or bar code that could be related to a database somewhere. The actual processing of the media was dependent on humans managing physical machines that read the tapes and connected various hardware devices for processing. The concept of dynamic reconfiguration of workflows was pretty rare. It was the era when we were still fighting physics and silicon technology was the limitation.

Fast forward by a decade and most workflows involved digitized files. Yet, the workflows themselves were predominantly replicas of the file-based workflows and databases full of records about media were predominantly used for searching, locating, and feeding the business pipelines of media organizations. We copied our manual processes into machine-driven processes. Although there was a lot of metadata in use, its interpretation was still very much dependent on humans reading the metadata and taking action or learning from it.

With recent developments and standards, we are entering an era where the metadata is now being used to drive and automate workflows. There is an implicit assumption that the metadata is correct and that concrete workflow decisions can be based on metadata and not interpreted via human operators. This new, high efficiency way of working can deliver great commercial benefit, but only if the metadata is good. So, what is good metadata and when does it go wrong?

Metadata – the good, the bad, and the ugly! Traditional metadata has often been split crudely into *technical* and *descriptive*. This crude delineation is no longer helpful because it tells us nothing about what is creating or consuming the metadata, or whether the metadata itself is reliable. Both types of metadata can be used for descriptive purposes or as an operational trigger or tool in file-based workflows driven by business logic.

Good metadata is typically immutable (can never be changed), accurate, and specific. Take, for example, some of the technical metadata in an SMPTE Material eXchange Format (MXF) file. We don't just say *"the video codec format is JPEG2000,"* we take great care to say exactly what sort of JPEG2000 and how it's been wrapped in MXF. This specific and accurate metadata is generally made by software and consumed by software.

Bad metadata is typically vague, and often woefully inaccurate. In an archive migration project that I had an opportunity to oversee, there was a lot of metadata in the database but only a few key fields had been historically used to drive workflows. The rest of the name-value fields in the database were simply annotations for a human reader. One particular field was frequently used to indicate a portion of the timeline that would be removed for a compliance edit. This annotation would be great metadata to improve automation, if only the field were used in a reliable way. Regretfully, in this particular library, the database field sometimes contained a timecode reference, sometimes a frame count and other times, a reference with no discernable use to anyone on the current operational team! Unfortunately, some key metadata was missing – what territory has this edit been made to support? This "meta-metadata" would be required to get the best automation out of the system. In the end, human interaction was required to validate the field and the cost of doing this additional validation step meant that the database field was never used. Management could not support the additional expense.

Ugly metadata can start off looking good, but then turn bad. A recent example was in an SMPTE Interoperable Master Format (IMF) workflow. The request was to have the name of the producer stored in the IMF Composition as embedded metadata. The local system designer applied sensible rules from his perspective, *"There is only one producer for the title, so obviously storing the name of the producer in a static metadata area is easiest."* No problem was discovered until much later when a promotional short was made with content from three different producers. The metadata that was previously considered static will now only apply to a portion of the timeline. Historically, this morphing of metadata type was never a problem because there was always a human to analyze the notes and "figure it out," but as we more and more automate our workflows with software, good timeline models for the data make life much easier and make the metadata immensely more useful. To future proof our databases, the initial data model design may seem ugly and unnecessarily complicated.

Mighty morphing metadata: During a metadata experiment conducted to design the descriptive metadata for MXF, we asked a group of German Public Service Broadcaster to choose their *must have* metadata set from a collection of 300 terms. We expected a basic unity of responses because when you view the collection of German Public Service Broadcasters from an international perspective, they all have very similar habits and models when compared to North American or Asian broadcasters. It came as a shock that each broadcaster chose a different set of *best* metadata elements. Many years later the moral of the story was discovered. Each of the broadcasters is fiercely protective of their own regional uniqueness and business model. The metadata chosen reflected that distinction. For this reason, in-house metadata models will nearly always be different and dynamic. The metadata must reflect the unique business if it is to drive the business. We therefore need to look at what is involved when exchanging metadata between different models.

A simple example to demonstrate the complexity of this situation is described as the "Address Book problem." Many data models for timeline media include the idea of roles such as producer, director, grip, etc. These roles tend to be fulfilled by people. An address book of people tends to be ordered by the person's name and will often contain a field that has their title. In a simple world, linking these two sets of data together should be an easy, static mapping. In the real world, however, there may be many such roles that apply to a particular title, and a person may have had several different roles. This many-to-many mapping problem requires active code and some careful logic to work in a reliable and automated way.

This morphing of data models is key to being able to interchange metadata between systems in a reliable, repeatable, and predictable way. Very often the interchange format will be different to the native format in either system. This helps improve test ability and interoperability of the metadata.

There are two leading methods for defining and representing metadata: XML vs JSON. Examples of these two metadata formats are supplied in the following Figure 5.4:

Many software developers treat the XML vs JSON question as a religious cause and will argue until someone buys a round in the pub over which is best. Pragmatically, they have different strengths and weaknesses. Here's my personal list:

XML	JSON
Strict methodology for defining data types, data relationships, and namespaces	Flexible, ad-hoc data based on numbers, strings, arrays, and generic objects
General purpose tools can validate an XML document using the right static schema	All the validation is done in custom code (but there are some good libraries)
Human readable but painful to do so	Very human readable
Great for rigid datasets such as interchange standards	Great for ad-hoc datasets such as the return values from REST API calls
Takes a while to learn, especially for namespace rich documents	Quick to learn
Difficult to make invalid documents if validating tools are used during creation	Easy to make invalid documents unless you are strict about the document creation

In the media world, we have traditionally used mostly XML and will continue to do so because software systems will need to exchange strict, complex data models without different design teams ever interacting personally. In the future, our workflow systems should be prepared to see more JSON appearing (and maybe some YAML too – "YAML Ain't Mark Up Language").

Metadata interchange formats are on-going in many parts of the world. Many expert industry teams have worked very hard to develop metadata interchange models. Some examples you may have heard of are the DublinCore, EBUCore, SMPTECore, and the SMPTE Broadcast eXchange Format or BXF. Each of these has a slightly different application space. It's also worth understanding that the MXF specification is just a metadata model for

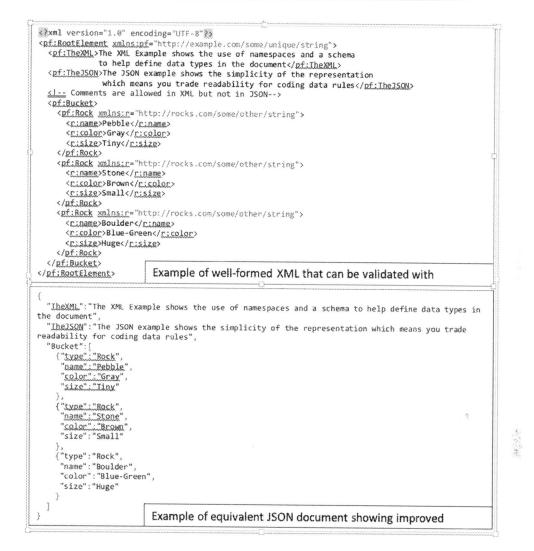

```
<?xml version="1.0" encoding="UTF-8"?>
<pf:RootElement xmlns:pf="http://example.com/some/unique/string">
   <pf:TheXML>The XML Example shows the use of namespaces and a schema
             to help define data types in the document</pf:TheXML>
   <pf:TheJSON>The JSON example shows the simplicity of the representation
             which means you trade readability for coding data rules</pf:TheJSON>
   <!-- Comments are allowed in XML but not in JSON-->
   <pf:Bucket>
      <pf:Rock xmlns:r="http://rocks.com/some/other/string">
         <r:name>Pebble</r:name>
         <r:color>Gray</r:color>
         <r:size>Tiny</r:size>
      </pf:Rock>
      <pf:Rock xmlns:r="http://rocks.com/some/other/string">
         <r:name>Stone</r:name>
         <r:color>Brown</r:color>
         <r:size>Small</r:size>
      </pf:Rock>
      <pf:Rock xmlns:r="http://rocks.com/some/other/string">
         <r:name>Boulder</r:name>
         <r:color>Blue-Green</r:color>
         <r:size>Huge</r:size>
      </pf:Rock>
   </pf:Bucket>
</pf:RootElement>
```

Example of well-formed XML that can be validated with

```
{
   "TheXML":"The XML Example shows the use of namespaces and a schema to help define data types in
the document",
   "TheJSON":"The JSON example shows the simplicity of the representation which means you trade
readability for coding data rules",
   "Bucket":[
      {"type":"Rock",
       "name":"Pebble",
       "color":"Gray",
       "size":"Tiny"
      },
      {"type":"Rock",
       "name":"Stone",
       "color":"Brown",
       "size":"Small"
      },
      {"type":"Rock",
       "name":"Boulder",
       "color":"Blue-Green",
       "size":"Huge"
      }
   ]
}
```

Example of equivalent JSON document showing improved

FIGURE 5.4 XML Example

representing file-based media and the IMF specification is a metadata model for handling and managing mastering workflows. Both can be fully expressed in XML. For more in-depth information, search Google for regxml.

We have all this metadata, where do I stick it? Now that you know how to represent your metadata and the pitfalls of different representations, you usually end up with the tough decision of putting the metadata in with the media or in a separate associated package called a "sidecar." There is a fairly simple test which nearly always works to determine if the metadata should be embedded in the media file or separated into a sidecar: Ask yourself this question, "If I wait several years and then sell the media to someone else, does the metadata need to change?" If the answer is yes, then the metadata should go in a sidecar. If the answer is no, then it's safe to put the metadata in the media file.

BUSINESS SYSTEMS

DAM/MAM Systems

Turning attention to business systems, the libraries for the media assets and their associated metadata are our first concern. As a newcomer to the industry, the DAM and MAM jargon seem to be used interchangeably, although they actually have very different applications.

Digital Asset Management is a document library focused on controlling, securing, finding, and leveraging content across an organization. DAM systems are typically used to enforce company governance, managing inter- and intra-office communications and templates, while making assets available to groups and individuals based on security clearance. DAM systems manage digital rights, enforce metadata and taxonomy, and optimize business processes, especially the interaction between individuals and departments, such as marketing communication, illustration, and graphics projects. DAM systems offer particular operational features, such as the ability to "check out" a project file and remove it from the library like a book from the shelf during editing or processing, and then "check in" the file when it is returned for general use. DAM libraries often manage complicated review and approval processes, mapped to the business logic of company requirements. A subset of the DAM system is the PIM, Product Information Manager, dedicated to the marketing department as an assistance to displaying the proper message to the correct channel in a timely and management-approved method. In addition to providing marketing communications, PIM systems manage data import and data model formatting and often interface to a customer facing "portal." Portals can be used for many different purposes, such as customer support, distributor communications and management, as well as media exchange between partners.

DAM systems can manage video and audio assets as a part of their library, but their tools and workflows are directed across all the internal assets and are primarily designed for managing internal and external written communications or training media. Media Asset Managers (MAM) on the other hand are often paired with DAM systems to extend their functionality to address media-specific chores, such as multiple audio translations, technical versions such as standard definition, high definition and ultra-high definition media, subtitles, edited content versions, and distribution versions. MAM systems employ internal playback tools with more professional video and audio features, extended data models to cover the many industry-specific data resources, and workflows dedicated to media and entertainment business chores. MAM systems usually include hierarchical storage manager tools to control massive amounts of storage in tiers as the media is often in huge files, and by design are more complicated and video specific in their focus.

So, a MAM is a MAM is a MAM, right? As Shakespeare says, "A rose by any name would smell as sweet," but in today's Media and Entertainment Industry, it is important to recognize that all MAMs are not the same, and while the language the marketers use to promote the products is often rich, expansive, and all encompassing, broadcasters and media professionals can save themselves a lot of pain, time, and money by understanding the various types of MAMs and workflow tools available, and where a particular product provides the best solution.

Some MAMs are optimized for linear play-out: Often utilized for enhancing or extending a Master Control automation system, these dedicated MAMs support a linear channel and effectively automate a number of repetitive station tasks such as dub lists and preparing media for air. They are architected to be workhorses and have solid integrations with the local automation and traffic systems. The "best of breed" systems employ the SMPTE BXF interface for this communication channel and come with ready third-party tool integrated workflows to handle the standard chores a Television broadcast facility addresses on a daily basis. There are a number of these cost-effective systems available, and typically their drawback is a lack of flexibility, business analysis options, or limits on their upgrade applications.

Traffic Extended MAMs: A subset of the MAM for linear playout operations are library management tools that are sold as embedded solutions or packaged with Traffic and Billing Systems, Content Management Software for OTT or VOD distribution, or Work Order systems. These extend a particular tool within an operation, and often suffer from the limits of the playout MAMs, as well as the limits of some vendors to devote development and support resources to more than their core product line.

Production Asset Managers (PAM): Years ago, we would call them PAMs, but vendors have tried to blur the lines of their products' abilities to widen their market reach. Library systems that prepare production schedules and manage workflows to create versions of media, track version relationships, and provide automated or manual quality control focused on a particular editing platform meet this essential need. Managing editing projects, "parent media," and all various "child" versions is of prime importance to production support software. These systems are often maximized to a specific editing toolset and have limited expansion capabilities, but what they do, they do extremely well.

Business Process Managers (BPM): Marrying your library to workflow engines for a network operation can bring some solid benefits, especially in task management for employees, automation of repetitive tasks, and business analysis for the executive team. These systems are more flexible, customizable, and provide more insight into large operations with reports and dashboards. BPM MAMs often integrate into traffic and work order systems, all internal systems including custom-built products, and external systems for supplier input and distribution capabilities. If your organization is looking to measure and manage increasing throughput and add new operations or functionalities, a BPM orchestration system may be the best solution for you. If your goal is managing your employees better, understanding their workload and watching the bottom line, BPM MAMs can provide the task management by individual and user groups to control all segments of your company. Key to finding a good BPM MAM is their support third-party vendor list. How many common products do they already support? These systems are usually customized to a particular operation, and they can be expensive to build due to the unique requirements of a particular broadcaster.

While many vendors will promote their software as an "enterprise" solution, only a very few actually have a system designed to support multi-site, multi-department, multi-tenant operations. Enterprise solutions build on BPM MAM tools and add the ability to limitlessly scale throughput, manage millions of assets, workflows and users, and support unique, geographically dispersed operations. PAMs and specialized MAMs are often built into these solutions. These are expensive, multi-year projects with major benefits. The resulting automation of workflows, oversight on task management and deep business analytics provide measurable reduction in overhead and labor costs.

There are also specialized MAMs built as sports and live events logging tools, archive management systems, distributed production applications for TV Group operations, native automatically scaling cloud-based systems, systems with internal programmable playlists – there are a number of specialized options available, each with particular benefits that may meet a defined organizational requirement. Yet, specialization comes with a price, usually a limitation in one or more areas of the product.

There are fundamentals of MAM technology that support automated and efficient workflows and are found in most, if not all, of the MAM products available on the market today. Architecturally speaking, most modern systems supply a browser-based user interface which allows quick deployment of updates and the ability to "skin" the interface in a user's language or save user-specific screen controls. The system should be ready to be virtualized, in other words the underlying architecture is ready to deploy on-premises, or in cloud infrastructure or in a hybrid combination of both. Frankly, in today's world with the proliferation of cloud service providers, an all cloud infrastructure is an acceptable alternative to a flexible architecture. Most systems are prepared with active Cluster design options and disaster recovery features. Object relational grid databases provide a fundamental library design that will allow true scaling of your metadata and the system and leverage the functionality of modern database software such as the rich reporting tools now available.

The core of a MAM is its search engine, the system that quickly seeks and displays the media according to user requests. Most companies have adopted "google-like" search tools that automatically offer suggestions as the user types, drawing on an index of search criteria. More recently vendors have added support for faceted search and "elastic search." The Graphical User Interface (GUI) can define special search criteria across millions of assets, often employing Boolean logic and wild cards, and many systems offer the user the ability to save unique searches for use at a later time. Some search engines have been enhanced with departmentalization strategies for assets, which is of great importance if the operation will manage millions of assets and associated files. Asset storage can be arranged in logical or physical storage, and systems can control media access with security measures.

The primary function of the MAM system is to bring media into the library in a controlled manner, allow management and enrichment of the assets by the library users, and then publish or distribute that media in many different ways in order to increase the monetization of the materials. The incoming process is called Ingest, and these capture workflows can be quite simple or multi-stepped with branching business logic to automate all possible acceptance of media deliveries. Typically ingest workflows will generate a reference "proxy" file, usually a frame accurate representation to the original media that allows annotations and edit decision lists to be applied by a user to the original media. Some systems will normalize incoming media to a house standard video format, but many media executives prefer to keep the incoming asset in its most pure form, and they save the original file as it was received, using the proxy to drive editing and versioning workflows.

Enriching media assets in a MAM includes adding metadata from multiple sources. Artificial Intelligence solutions can automatically annotate every frame of a file or program, convert the audio tracks into scripts, automatically mark points of interest such as product placements and on-screen graphics, etc. Automated quality control systems can annotate the proxy file with technical and visual data based on the testing criteria. Descriptive metadata can be translated into many languages and added to the database referencing to the original media file. Audio versions, such as stereo pairs, 5.1 surround sound and Dolby ATMOS, and foreign language translations can be associated with the original media, and the video can be marked for edits to align to the new translations as well as subtitles. High Dynamic Resolution (HDR) technical data can be added to the media asset and referenced in the MAM. Unique business requirements drive the enrichment process, and the MAM is receptacle for these enhancements and the maintenance of their connections to the original media files.

Publishing, or export of the assets from the library, is another fundamental function of the MAM. Leveraging integrations with third-party tools such as watermarking software, Digital

Rights Management, video format transcoders, etc., allows MAM technology to maximize the numbers of versions the assets can take upon export, and most of these functions are automatable in current technology. The publish function of MAMs has been an industry focus for years, driving innovation to reduce labor and increase speed and efficiency for servicing social media, streaming services, video-on-demand and over-the-top services. Publishing can be as simple as a "drag-and-drop" of a file to a new folder, or as complicated as managing thousands of profiles in third-party tools.

In the newest technological advances, cloud-based tools designed as microservices in dockerized modules permit the scaling up and down of operations that allows cost management for unpredictable workloads. This is a boon for start-up businesses or companies that want to "pop-up" a channel or deploy an increase in capacity for a short-term production project. These MAM systems often tout their "native" cloud operation and link with a particular public cloud service provider. Recently there have been some highly publicized network projects to move MAM operations to the cloud to support cost effective, and rapid deployment of linear channel broadcasts. Cloud infrastructure offers MAM systems new flexibility and scalability.

Media has been stored and prepared by the MAM, and in some cases that media is even made "rights aware" by connecting it to the program and rights management tools. This is the next process that media workflows address.

Programming Workflow

Before we discuss the programming workflow, it would make sense to define Rights and Rights management. The rights to programming seem simple to understand, but in reality, rights are difficult to manage. There are two types of "rights," rights for material that your company owns or manages and rights for acquired content. An easy way to describe these two types are as "rights in" or media acquired by your company for use, and "rights out" for product you own and distribute.

"Rights out" management systems are of key importance to asset owners and are often directly connected to an enterprise MAM to control the library access or manage/limit distribution. They track the media asset and its constituent parts and elements, as well as the contractual obligations the company may have in regard to those components. For example, music may require ASCAP and BMI license payments, or be restricted in some regions. Actors, directors, and writers' guilds, etc., require percentage payments based on the distribution receipts. Distribution management includes coverage of territories and transmission format such as over the air broadcasts, IP streams, On-Demand file distribution, watermarking requirements, and Digital rights management encryption. These software systems manage the legal distribution contracts and support financial tools. "Rights out" workflows are typically managed as MAM extensions in conjunction with Content Management Systems (CMS) for packaging, and these operations will be discussed later in this chapter in section on "Automating Workflows."

For an analysis of the workflows for "Rights in," acquired content systems and the typical workflows associated with them, **Peter Storer**, Founder and CEO of the industry leading STORER TV Program Management software, has supplied a detailed description:

Programming, or for many in the industry simply "content," covers everything from local live productions, like news and sports productions, to syndicated content purchased from large content distributors, such as *"Wheel of Fortune"* from King World Productions, Inc. Networks, like, CBS, NBC, ABC, and Fox act as both creator and producer as well as the purchaser of lots of content. In return, they license the use of this content through affiliation agreements to their local stations. Cable Networks also both produce and purchase lots of content in order to fill their 168 hours of programming every week or 8,736 hours every year. Based on a 2014 report by Nielsen, the typical American has access to almost 200 channels of content of which they watch on average only 17 channels.[1] Think about what it takes to manage that 1.7 million hours of mostly unduplicated content every year.

For every hour of non-live content, there is often hundreds of hours of video/audio captured via various devices that must be reviewed, color corrected, audio enhanced, edited, close-captioned, and then delivered to your viewing device, which also has evolved significantly from the days of the standard-definition TV. In short, the process, the workflows, are staggeringly complex and diverse depending on what you are recording, how it will be shown, and who will be watching it.

Let's consider a relatively simple workflow, say for a live news program. This program airs every day, at the same time slot for exactly the same length of time with the same format or structure (more on this later). Out of the 30 minutes total, you subtract out the number of commercials, let's say 8 1/2 minutes along with 10 seconds of promotional material, and you are left with exactly 21:20 minutes of time to fill every show. You plan on 3 minutes for your standup weather, 4 minutes of sports (these might vary significantly depending on the market), and 20 seconds for your open and closing graphics and introduction, and that leaves you with 14 minutes for actual news stories. As a result, you might need fourteen 1-minute stories, or five 1-minute pieces and three 3-minute pieces, or some combination that fills the time. Your newsroom assignment editor hands out the stories to their reporters based on what they think will happen that day. In addition, there are also a number of "timeless" pieces that can be used as filler. All of this comes together as the day progresses with last minute changes until airtime. When the final pieces have been chosen, their order decided, the program's director cues for the program to start. It is a time compressed, intensive, and unforgiving process that happens every day to deliver the news, yet the workflow is pretty straight forward and well established.

Separate from linear television and network broadcast channels is the News gathering workflows which constantly adapt to changing technology and vendor innovation. Over the last several years, technology has evolved to make the tagging and movement of video simpler and today it is practical to do live feeds from the front lines of virtually any story direct to the station and then on to the viewer. These workflows are defined by the key vendors in the space and are quite competitive in features and operations – news gathering organizations should investigate the product defined workflows from each vendor and source the model that best suits their business operation.

Turning back to the more traditional content workflows and how a television station or cable network decides what to put on the air every day and the process of actually making it happen. The first step is the same whether you are writing a book, giving a speech, or making any type of presentation: It is to know your audience. For a traditional broadcaster, there is a long history of surveying their audience using research companies like Nielsen, Arbitron, and others that provided a customized look into who, when, and why people watch what they do. These preferences are why the news is on when it is, and network programming starts when it does. Over any particular day, the TV audience changes its demographic makeup, minute by minute, but interestingly, when averaged out, the actual viewer levels stay pretty stable for any particular time period. Hence the decision to place any particular program in your inventory will be influenced by the demographics that you determine are available at that time. It will also be influenced by the competitive content already available in the timeslot. You might own a really strong game show that targets people 35+ in age, but if there are already three game shows airing in the marketplace, you might consider counter-programing them with news or alternative series content and placing the game show elsewhere. In short, what you put on in each slot is a very complex decision process that many consider more art than science.

In fact, it is a little bit of both. You have to consider what you have purchased or otherwise have available, what it costs to air in the time period, what it may generate in terms of audience performance and hence its attractiveness to your sales staff to be able to sell advertising in and around it. If a show costs $10,000 to air each day and commercials will sell for $1,000 per 30 second and if there are 6 1/2 minutes of time to sell based on the shows format, then your profit will be $3,000 each day. If you can only average $500 per 30, then you will be losing $3,500 each time it airs.

Of course, you sell advertising on a future prediction that it will perform at a certain level ("an estimate") and what it actually delivers in audience is only known once it starts airing. Hence the reason networks often produce lots of short-run series (7–8 episodes) then run them and see how the audience likes them. Poor performance results in quick cancelations.

Once content has been selected, acquired, and scheduled to air, you often also have to arrange for delivery to your location. This process can range from standard mail delivery of an analog tape to digital files being streamed via private satellite hookups. Once your content arrives, it will often need to be ingested into your MAM, and sometimes it must be transcoded from the production format into the format you have standardized for linear playout within your own operational environment. As part of this process, it may also need to be viewed by someone to determine that it will play back correctly and that there are no objectionable elements based on your predetermined standards. Once approved for air, the show may need some editing to prepare it properly. There is a huge range of options depending on whether the content will air with or without commercial interruptions or many other considerations. Building in appropriate fades to black at designated transition points often need to be done and, if there is any objectionable material, this may also have to be edited out. The final edited version is given an identification code, often referred to as a "house number," that is then used to reference its use by all the operational in-house systems: typically, the programing software, the traffic software, the MAM, and Master Control automation systems.

Once an asset is selected to air, the house number will ultimately appear on an automation log that controls the order of playback. The structure of the program (the exact timing of each of the program's segments between the black fade points) must also be entered so the automation system knows exactly how long each of the segments will play before they switch to other content. As the content airs, a record of its exact playback is maintained in order to determine if anything aired improperly. More important, these records act as a proof of airing for commercials, and also serve as an accurate record of what each station aired during every day to comply with any Federal Government investigations. The program management system will often then be updated to reflect that the specific content has been aired and this may be used for various financial reports. As a result of it being aired, the cost of that program has been incurred and its remaining value has been reduced or amortized. If the contract for the use of the content indicates a limited number of airings, the use of the program will also be reduced by one. A history of each airing of every piece of content owned or produced by the station is also typically maintained in the program management system.

This discussion concentrates mostly on content workflows that are used within a traditional TV ecosystem. Today, an alternate content delivery process overshadows this linear broadcasting methodology by delivering huge amounts of video content via the Internet. YouTube, Hulu, Netflix, and the hundreds of other Over-the-Top content sources provide both traditional series, feature and short length films, as well as thousands of other different formats that are consumed by today's viewer in so many ways. Much of this content does not follow the traditional Network or TV "workflow" process, but it has become a significant challenger to the traditional video content delivery systems.

Media Planning and Promotion in Linear Network and Broadcast Television

Assets are scheduled but no one will watch them if they don't know about their broadcast. For understanding the next steps media takes as it works through the operations of a linear broadcast, we turn to the President/CEO of Effective Media Systems, the supplier of the industry leading GRIP IT! Software, **Brendan Kehoe**.

Success in Television should often be viewed as a three-legged stool. Acquiring or making compelling Programs can be viewed as the first leg. Without compelling Programs that attract and keep your viewer interested, and engage them in the plot line, characters, and the outcome, success will be fleeting. The second leg of the stool is to make Promo materials, those 60, 30, 15 second clips that outline what you are offering as a product to the viewer, and interest them in "purchasing the product" – getting them to tune in long enough to be willing to sample and come back again for more.

But having those two legs of the stool is not enough, unless you can get those Promos to the viewers most interested in that type of Genre/Program, and get to those viewers enough times such that what you are offering is "TOP OF MIND" – they remember your offering, and when it is being offered to tune in.

That is the art of Media Planning and Promotion Scheduling. This art is no different than what advertisers spend so much time and effort in planning, executing, and validating results of their own commercials on all channels. Channels need to treat their own advertising with as much seriousness and planning as those who are willing to pay to advertise their products on your channel.

If you fail at this art for yourself, advertisers will not see the ratings success they expect in your programs, and your revenue stream will never meet expectations. There have been many wonderful programs that have been produced at great cost but have failed because this last leg of the stool (and the second leg of the stool) was executed poorly. There have been many a "why did we fail meeting" on any channel, where audience has been polled on why they did not watch, and often the results are:

1   The Promos I saw did not really tell me what the show was about, or did not interest me

2   I'm interested in only Sports or News, so why show me a promo for a Soap Opera on in the afternoon

3   What Program – never saw one but wish I did

Well executed Media Planning and Promotion can overcome this. In fact, well executed Media Planning can even get viewers to tune in to a terrible program, at least for the first few minutes.

So, what is involved in this art? It involves research, some industry math, and an understanding of what drives viewers. Let's discuss these at a high level.

Being the Bridge: As someone engaged in the role of Media Planning, the first thing to recognize is you must be the bridge between many groups at your channel. Programming will have their own ideas about how to promote a program, Research will weigh in on what they think may be successful, and Advertising Sales will weigh in on what they hope will happen, and what they hope to sell the Program at and to what advertisers. The Media Planner will have to take all these opinions into account, and yet be the singular entity to understand everything else going on at the proposed launch time and push back where necessary.

This involves understanding the "Media Math" of your channel and audience. Much like your Mother may have told you to first eat your vegetables off your plate, lets deal with the math part first of Media Planning. The key terms of this are: GRPs or TRPs, Gross Ratings Points or Target Ratings Points. When a promo spot airs, each spot will reach a certain number of viewers. In each demographic, the numbers will vary greatly. In a sporting event, you may reach more Men 18–49 (M1849) than Women 18–49. In a reality program (think "the Housewives of…") it will be the reverse. For each Demo, a number will be assigned by the reality of who watched. This number is expressed as a percentage of that Demo audience in your "Universe." A 1.0 Rating for the demo will count as one GRP or TRP for that demo. The raw count is called Impressions. All Media Planning is done only by GRP/TRP achieved. As well as Reach and Frequency (R&F).

Many times, both Programming and Senior Management will ask "how many spots ran?" The job of the Media Planner is to re-direct these questions to "how many viewers did we reach?" or as you see in a bit "How many viewers did we reach enough times to get them to tune in?"

The "number of spots ran" question is immaterial. A spot in Prime Time may reach 600,000 viewers. A spot at 3AM may only reach 10,000 viewers. So how does the number of spots run make a difference?

Reach and Frequency (R&F) is important: Reach is defined as "How many of my audience have I reached?" You can never hope to reach 100% of them with limited Promo Time and current viewing patterns of viewers. To reach all viewers who are in your Universe, you would need to run a 30-second spot, 24×7 in every break position to achieve that. Unfortunately, many viewers will have abandoned your channel out of being fed up seeing the same spot so many times.

The goal of any media plan should be to reach as many of the target audience as possible, without entering "negative frequency." Negative frequency is the point where viewers have seen your promo so many times that when it comes on, they leave your channel, to ill effects. Starting out a Media Plan will require at least picking a Demographic (Demo – M1849) to which you hope to achieve effective frequency.

Typically, a good media plan will try and reach approximately 70% of the Universe of viewers who WATCH your channel, and to reach them enough times in a "purchase cycle" to influence their viewing behavior. This is called Effective Frequency. That number will typically be between a minimum of three, and a maximum of seven. So, most Media/promo plans will look to achieve this over a period of four weeks, four weeks considered the Buy cycle for new shows.

There are systems that will project these numbers for you, as it requires the ingest of huge volumes of research data to make these calculations. These measurements use Audience Targeting/Affinity. Just getting to an effective frequency against a specific demo is usually not enough. A Media Plan will need to target specific groups of viewers within a demographic. For instance, if you are launching a Situation Comedy, you want to reach those viewers in the demo that view Comedy programs, on either your own channel, or on competitor channels.

There is research data that allows for systems building a Media Plan, with your input targeting information, that will track viewers in that category, and will direct your promo spots to where you can reach those exact groupings effectively, without wastage.

As well, those systems need to understand when the program is running, taking into account the viewers that it will try to reach are even available at that day/time to view the program.

A critical consideration is Spot Length: The Media Planner also needs to take into account the length of the Promo spots available, and when they should be run. Promo spots in the beginning of the Media Plan execution should be long enough to tell the story of what you are promoting to attract viewers. This usually requires a spot of at least 30 or 60 seconds. Many Media planners make the mistake of running shorter spots (15) at the onset of the Promo Campaign and scheduling the longer length spots closer to air.

Audience testing has proven that initially running the longer spot at the start of the campaign, then using the shorter "cut-down" of that spot closer to premier, causes viewers who have seen the longer spot first to immediately recall that when they view the shorter length spot. It's the way the human mind works. Have you ever picked up a book, read the first two pages and realize how it ends because you read it years ago? Promotion spots have the same effect.

There are added tools available to the Media Planner in reaching audience. Especially for reinforcing the messages they received earlier. These are the in-show graphics you see run on many channels. They go by many terms – Lower thirds, Snipes, Violators, Ghost Bugs – but serve a very distinct purpose for audience targeting.

There is a great deal of clutter on most channels. Break durations seem longer every year and have more commercials in them (especially since the advent of the 15 second commercial). Your Promo spot runs the risk of being lost in this clutter. An effective mechanism is to use these in show graphics as "reminders." In the show segment following a break where your Promo spot ran, scheduling a 12-second graphic as a "reminder" – Premiers Tuesday at 10PM – will assist in getting the target audience to remember what you are offering. They also help reach viewers who may not watch any breaks at all. The planning of these should be just as sophisticated as the rest of the Media Plan.

Your own Promotions, when added up, make you the single biggest advertiser on your channel. The use of these tools will help you generate success. Many a failed, expensive program has resulted when effective Promo Planning is skipped.

Sales/Traffic/Scheduling

Karyn Reid, the Vice President for Broadcast Systems at Fox Corporation, has been a Product Manager at both broadcast traffic/automation and program management software vendors and has been involved in the SMPTE BXF schema committee since its inception. In this section she provides insight into the systems that monetize the program channels, the advertising sales and scheduling tools.

Ad Sales and Traffic systems have taken different evolutionary paths depending on the geographies and businesses that are supported by the software design. Different terminology has also been adopted to refer to similar processes, depending on the business or region. The system that modules are contained within can also vary, so for this review we will focus on the functional module versus the system that contains them. Most of these systems now include integration between their elements, and much of the data that in the past was manually entered in multiple systems can be entered once and flow through the systems at the required time.

Ad Sales

Ad Sales systems typically contain modules that support advertising campaign planning, proposal creation, maintenance, and reconciliation.

In the United States, ad campaigns are typically referred to as Contracts (or Orders) in Local TV/MVPD sales and Deals in network sales. Network buys are typically negotiated using a Cost-Per-Thousand (CPM) currency, while local buys use Cost-Per-(Rating) Point (CPP). Deals or Contracts are typically planned and proposed to buyers, and once accepted create commercial "spots" that are scheduled on the Traffic log. As the spots are aired, there is a matching process between spots scheduling parameters and how they actually were broadcast, referred to as Reconciliation (Recon). The Recon process is usually contained in the Traffic module and when complete, the Ad Sales module is updated with the actual airing time.

There are two workflows for spots once Recon is complete: one to manage spots that have not aired according to the Contract parameters, and a second that compares the estimated viewer ratings measurement against the actual measurement. If a spot does not air, or does not air as scheduled, then a "Makegood" or "Re-Expression" is required to resolve the discrepancy (the second spot "makes good" the missed spot or spots). This can be another spot in the same program or time period or another spot or spots in a program or time period with similar viewer demographics to the originally purchased spot. This second process is called Stewardship or Posting. Once the audience delivery data is received and processed by the seller, the campaign delivery is compared to the estimated delivery. If there is a shortfall in delivery, then additional spots are scheduled at no charge to the client to make up the deficiency.

In other regions, campaigns can be sold on an impression basis, which is much more like digital advertising. Buyers are sold a number of impressions, and as campaigns air and audience measurements are received, the campaigns and/or spot placements are modified to assure the promised impressions are delivered.

There is also a move to include digital campaigns within linear campaigns, though there are limitations on the extent of the implementation, especially around the electronic billing function for the omni-platform campaigns.

CONTENT MODULE

Traffic Content modules are a repository for limited content metadata and primarily support short-form content such as commercials, promos, and IDs. Secondary Event content can also be managed though the scheduling parameters are typically managed in the log module.

Some Traffic systems can also manage program content metadata, while in some regions this is typically maintained in Program Management and Rights systems. Some systems are now supporting frame level time content metadata and can import technical metadata from MAM/DAM and Playout/Automation systems.

TRAFFIC/COPY INSTRUCTIONS

While the Ad Sales system manages the request for airtime, what content is requested in that time is managed by a Traffic or Copy Instruction module. Many Traffic Instructions are simple, such as rotating two pieces of commercial content equally for the duration of the campaign. Others can be extremely complex, such as defining a detailed rotation pattern or changing what content airs by day, and/or by time period. For Networks the commercial content can be defined on a spot-by-spot basis. Traffic Instructions are also often very volatile and can be revised by the buyer multiple times during a campaign. Examples of frequently revised commercial content include movie trailers, and in the US, political commercial content which can be changed multiple times a day.

LOG MODULE

The Log module is where all the details come together to create the schedule for a linear channel. Programs are scheduled well in advance and include the program structure (Format) including the expected number of program segments and commercial breaks. Formats can also include secondary event scheduling details for both technical commands and promotional or ad sales content. Spots can be scheduled in the breaks many months in advance of the actual log date and are moved and updated as more information is added to the Log. As the air date comes nearer, more details are added to the Log such as Program Episode segment timings and Secondary Event Detail, either by integrations between systems or modules or entered manually by Traffic staff. Then, the spot scheduling process consisting of scheduling, moving, and optimizing spot placement begins as new or revised ad campaigns are processed by the system. There is typically a point in the Log processing workflow were the Traffic staff lock out external processes, and manually complete the log. They ensure that all open time is filled and that all spots are scheduled in compliance with their campaign parameters. The Copy Instructions are applied to all spots, and any spots missing content are removed from the log and replaced with either another spot or by promotional or filler content. When the log is completed, then typically the playlist is created and sent to the Playout system. Changes that need to be made after this point can either be handled manually, or by a Live Log mechanism provided by the Traffic and Playout vendors.

SCHEDULING ENGINE

The process to schedule and optimized ad spots is a very complex algorithm and continually updates the Log as ad campaigns are entered and modified or Log events are revised. Scheduling Engines ensure spots are placed as details in the campaign, while also ensuring protection from competing advertisers or restricted content. The Scheduling Engine should also provide an even rotation of a campaign's spots across days and within a time period ordered. An example would be a campaign that ordered ten spots a week, ordered Monday–Friday, from 06:00 to 10:00. A good rotation would ensure the spots are evenly scheduled across all days ordered, and within the time period. An example of a bad rotation would be most spots being scheduled on Monday and Tuesdays, only most within the nine to ten hour. Spots are also given a priority within the Ad Sales module, and spots with a higher priority (and hopefully spot cost) will have a better chance of receiving a preferred schedule.

RECONCILIATION

Once a playlist or event has completed airing, the airing details are returned to one or more business system modules to be processed. The Reconciliation (Recon) process compares how an event was scheduled or ordered, and how it aired. If there is a significant variance, such as a spot not airing, or airing out of the time period requested, then the event is flagged and will require resolution. The Recon process is typically automated, with a manual evaluation of the events marked as out of requested parameters. Each business system can process the Recon details from the playout system, or a single system can process the output from the playout system and then send the relevant details to the other business systems.

INVOICING AND ACCOUNTS RECEIVABLE

After the Reconciliation process is completed, invoices are produced, typically on a monthly or weekly basis. These invoices can be sent to the buyer either electronically or in printed format. Many Traffic systems include an Accounts Receivable module though many networks use their General Ledger system instead.

REPORTING

Ad Sales, Traffic, and Scheduling systems include detailed reporting modules to track projected and actual revenue, commercial inventory available to be sold, and regulatory compliance. As Ad spots are scheduled and revised and Logs are edited there is an impact to the projected revenue and inventory, so these reports can be run during the business day to get the most current data.

Master Control Automation / Playback

With years of experience in broadcast operations and as VP/GM for iHeart radio's Florical Master Control Automation software division, **Shawn Maynard** *provides a thoughtful analysis of the next workflows that media follows on its way to transmission, Master Control Automation and playback:*

Since the beginning of television, various electronic machinery has been in use to produce the product that viewers have enjoyed for ages. As the industry evolved, the electronic machinery became more and more intelligent and eventually allowing external control. This external control, once used for more conveniently placed buttons for Master Control operators, birthed the dawn of automation. The inception of automation began with very basic capabilities to manage tape/reel machines for recording live content for later playback, then quickly evolving into full automated control that manages the entire Master Control workflow chain. It would be very difficult to operate a modern television Master Control environment without some level of automation governing the process. Automation moved from a luxury to a necessity for the future of television content management and playout.

Television automation can be divided into three basic functions within Master Control: Acquisition and Preparation, Traffic and Management, and On-Air Playout. Each of these various functions will be explored from a high-level point of view providing the reader with a general understanding of the role of automation within a broadcasting environment. Before diving into each individual area of function we must first provide an overall description of the workflow of Master Control to understand the environment that demands the need for automation.

A viewer is sitting on their sofa enjoying their favorite television program without any thought of how it was delivered to them in the full package in which they see. The fact that the process is invisible to the viewer is a good thing and a testament of the success of auto-mation. A broadcast company includes many different departments that are responsible for various aspects of the product that eventually is delivered to the viewer. From a very high level, a typical facility will have a Sales department that is responsible for selling commercial time (i.e., commercials/spots/ interstitial content), a Programming department for tailoring what "shows" will air, a Creating Services/ Marketing department for promoting the product to the consumer through various means, Production department for creating unique and live content, and a Traffic department that is responsible for aggregating all the business needs and create a schedule that contains the detailed instructions on how each element will play out for a given day.

The very end of this television "sausage factory" process chain that is responsible for what the viewer will see is Master Control. They are the tip of the funnel that receives, manages, and plays out the content that comes from all the various departments to fulfill the instruc-tions of the schedule from Traffic. The strongest relationship and acting liaison between Mas-ter Control and the other business units is Traffic. In today's automated environment the responsibility of how content will air has shifted from the Master Control operator to Traffic. The automation system precisely executes the schedule provided.

The first steps media takes in these workflows is in the acquisition and preparation phase. Traffic may be responsible for the schedule of instructions of how and when the precise content will play back but they do not generally provide the content itself. The responsibility of Master Control is to acquire the content needed to fulfill the Traffic schedule. In today's operating environment most content arrive via Internet delivery into a local storage device as electronic files with metadata instructions attached and, although rarer, some content arrives via satellite stream that must be recorded into a local storage device. Automation has evolved over the years to manage the entire acquisition chain by monitoring "watch folders" for the delivery of electronic files and matching the inherent metadata provided with the local copy instructions of the asset from Traffic.

The automation will automatically retrieve the file, run it through a transcoding process as necessary to create a localized version, update its database, and move the finalized copy into the on-air video playback system. Additionally, any satellite or live video feed content is handled in a similar way. The automation will control the various machinery needed for receiving the feed to convert the content into an electronic file for the process described above. In general, all of the machinery (equipment) necessary for acquisition of the content is managed and controlled by the automation system. The operator interfaces with the system through a GUI (Graphic User Interface) in order to verify or manipulate the content once it is acquired within the system.

The main responsibility of the Master Control operator is to verify that all content is in the system and ready for on-air playback. By using various applications provided by automation the operator will manage the content to verify the quality and ensure that it will playback as instructed within the Traffic schedule. For long-form programs (i.e., shows) the operator will verify that each segment matches the format that is scheduled. Most Internet-provided content will contain format metadata that the automation system will automatically incorporate within the on-air database and operators mainly use the automation GUIs to verify or make any necessary adjustments to the timing details of each segment. The ultimate goal is to ensure that all content is prepared and ready for on-air playback.

It the symbiotic relationship between Traffic and Master Control cannot be understated. It is imperative that these two departments "dance to the same rhythm" and work closely together. Master Control is at the mercy of the broadcast schedule and metadata instructions for interstitial content that is provided Traffic. Whereas Traffic is dealing with sales, programming, creative services, News, and advertising agency instructions, Master Control is utilizing electronic equipment to manage the essence ("physical representation") of the content itself. They have at their disposal all the equipment and software necessary to ensure the integrity of the on-air product.

Advancements in technology standards has significantly helped to facilitate the integrity of communications between these two departments such as SMPTE's ST-2021 (Broadcast eXchange Format). BXF has standardized the electronic communications between the Traffic software system and the Master Control automation system to ensure they are speaking a common "language." The growing use of BXF within television is helping to maintain metadata workflows from programming and commercial content that carry through the Traffic business system on to the on-air Master Control automation system.

Asset Management is also a large part of the overall responsibility of Master Control to maintain content inventory for playback. Storage architecture varies from one business to another based on the amount of content that can be accessible by the automation system. MAM (Media Asset Management) and DAM (Digital Asset Management) system integrations are becoming more prevalent and necessary as business requirements become more complex and move beyond simple playback instructions. The integration of these systems within automation has advanced the capabilities of the playback system significantly. The storage management can now be maintained in more efficient technology mediums like off premise solutions (cloud) or complex archive management solutions that are seamless to the on-air automation system.

As already detailed previously, the schedule instructions provided by the Traffic system are what the Master Control automation system utilizes to manage the on-air product. Those instructions account for every second of the 24-hour broadcast business day. Depending on the type of on-air product, the complexity of those scheduled instructions vary: for simple on-air products, it could be playing back pre-recorded content in a back-to-back linear fashion requiring very little interaction of Master Control operators outside of maintenance issue; for more complex operations a schedule may include a variation of live programming, pre-recorded content, graphics, and dynamic interaction (e.g., sports, breaking news, etc.).

Master Control automation should require little to no operator intervention to perform the duties found within the Traffic system. The instructions should be clear and concise for the automation system to fully execute them consistently and flawlessly. In a traditional Master Control environment, the automation system will be managing and controlling various equipment needed for playout.

This equipment includes but is not limited to a video router, Master Control switch, video servers, keyers, graphics systems, and more. More modern systems are using software versions of traditional hardware that run on COTS (Commercial-Off-The-Shelf) computers called Integrated Playout or "Channel-in-a-Box." In the early adoption stage of the always evolving Master Control automation is moving to more advanced technology in virtualized environments that move away from monolithic software applications to micro services that handle the functional requirements in a more efficient and economical workflow.

What does the future hold? It is impossible to accurately predict details of the future of Master Control but certainly the broadcast industry is moving quickly in the footsteps of the IT industry by using more software-centric solutions to provide greater quality that is more adaptable to an ever-changing consumer expectation.

There have been three main technology shifts over the past 15 years that has radically changed the landscape of Master Control. Moore's Law has remained faithful by accurately predicting computing capabilities that operate at speeds to enable more software-centric tools to meet and exceed hardware equivalents birthing the use of COTS computers in place of purpose-built hardware. The shift from SDI (Serial Digital Interface) video standards to a wider-industrial IP (Internet Protocol) standard allows more IT industry tools and deeper manipulation of the video workflow without damaging the integrity of the product. Finally, the expanding use of cloud technology has moved the product out of the purpose-built facility to a more virtual environment that can be controlled and managed throughout software solutions.

These shifts of technology demand more and more need for intelligent Master Control automation solutions that are reliable, flexible, and simple to integrate. The mantra of doing more with less is remaining true. To add to that, there is a desperate need for automation solutions to simplify the inherently complex environment of disparate technologies.

In conclusion, Master Control operations are like snowflakes; to the casual observer they are all the same. However, the devil is in the details. No two operations are ever exactly the same. They each have unique requirements based upon the on-air product they control. Automation is key to keep advancing the capabilities of the on-air product by continuing to make the process more and more simple to manage.

How BXF Can Help

Again, drawing on Karyn Reid for her expertise, here is a description of BXF and how it applies to our media workflows:

The SMPTE Broadcast eXchange Format (BXF) standard was originally planned as a way for business systems and playout vendors to reduce the number of software interfaces they were maintaining. Each time a vendor released a new update, each of the existing interfaces had to be regression tested to ensure that there would be no impact on TV stations' critical schedule functions. As the group began its work, it became clear that there were many other workflows that could be improved with an agreed upon standard, and the work also expanded to include transactional as well as file-based communication between systems.

Early in the adoption of BXF, there was a common misconception that BXF was a "plug-and-play" standard. It is a very expansive schema and a single transaction can be expressed in multiple ways, so it is key that all systems in a transaction agree on the structure of the data. BXF is also extensible, so any data missing in the schema can be added to an integration as long as the vendors agree to the schema expansion.

One commonly known application of BXF is Live Log, which allows changes made in a Traffic Log to be updated in Playout in real-time. Live Log replaces a very manual process for traffic schedule revisions made after a Log is completed. The benefit of Live Log is that changes can be made in the Log module, where the revision can be validated against ad sales, promotion scheduling, and Program Rights parameters, ensuring a late change will not violate any of the many constraints managed by these systems. A caution to this is that the Log module may not know the actual "live" time, and changes made too close to live, or even past the actual time, may impact the on-air product. Any implementation of Live Log should include at least one gating mechanism to ensure changes are made far enough in the future that issues, such as missing content, can be resolved before the event is scheduled to air.

BXF also supports Schedule and Program Format communication between Program Management and Traffic systems, Content updates including dubbing and purging commands, and Reconciliation (As-Run) data.

The ability to communicate real-time Recon data can have the greatest impact on a media organization's revenue. Traditionally, the Recon process is completed on the morning of the first business day after the schedule log was completed. If a spot was missed on the last day of an ad campaign, but the traffic staff was not notified until the next business day, the result can be a credit for the missed spot and lost revenue. If the Traffic system was made aware of the issue as soon as it occurs, the staff, or the system itself, could resolve the missed spot and save the revenue.

Being able to add or remove content records or update the status of content is another powerful feature of BXF. Many times, content is received via a Content Delivery Service, and then the metadata is manually entered in the business systems. This can result in keying errors or mismatches in metadata, which can flow through the various workflows to the invoice and delay payment. By using BXF to pass a single source of truth between systems in the media workflow, processes should be smoother with less confusion and error. Real-time content deletion can resolve a common issue, where content is deleted by Traffic staff and the Content ID is reused shortly afterwards. If the deletion of the original content is delayed in the playout system, the old content may air in place of the new content. The use of BXF to support Content IDs such as Ad-ID in place of legacy House Numbers can help reduce the risk of incorrect content being aired.

BXF has been used to pass inventory availabilities between Traffic and Promo Optimization systems, Promo schedules including secondary events, program schedules between Networks to Broadcast stations, and has supported Metadata communication.

As media workflows have expanded to include systems such as MAM/DAM, Content Delivery Systems, and Enterprise Service Buses (ESB), BXF has found a place in the communication workflow.

The BXF committee continues to refine and expand the schema which now includes Traffic/Copy Instructions and enhanced Secondary Event commands along with many small updates that support new technologies.

Quality Control

*Quality control in a broadcast operation is different from the QC operations used in post-production. To understand the workflows, we have tapped **Frans De Jong**, Senior Engineer, European Broadcasting Union to supply an understanding of its wide-ranging impact:*

Quality Control (QC) is essential for making sure content is technically useable and editorially correct. A lack of proper QC can translate itself in delays, viewer dissatisfaction, and potentially hefty costs for recommissioning and penalties.

The Desire for Automation

QC processes have developed slowly over many years, based on linear workflows that were unique to each content provider and often targeted a single output, such as traditional broadcasting. Much of the QC process was manual, relying on 'golden ears & eyes.' The move to file-based workflows combined with an increase in the number of outlets has led to a situation where the amount of content and versions is so large that automation is desired. It is also desired because a large part of the technical QC is abstract; human beings are good at verifying visual information, but not very good at sifting through gigabytes of an MXF file to spot a single byte error. Traditionally this was less of an issue, as the limited number of media supports (videocassettes) already guaranteed many of the technical parameters; putting a Betacam cassette in a DVCPro player was physically impossible.

QC Workflow

The amount of QC automation that can be achieved depends first of all on the types of checks a company wants to perform. Computers are good at (file) format checks, but humans are still better at spotting specific baseband video issues and, of course, at editorial QC. Practical workflows make use of this complementarity, typically starting with automated QC, followed by the manual checks. If a file can be failed fast automatically, it prevents wasting valuable human QC resources (Figure 5.5).

Example workflow consisting of an automated QC step followed by a manual one.

QC Delivery Specifications

Media workflows often span multiple departments, companies, or even countries and continents. This complicates QC, as it risks checks not to be performed or performed multiple times, costing time and money. Also, media facilities do not necessarily possess the wide and deep skillsets to perform all QC in-house. This can be addressed by clearly specifying what QC is expected as part of a media organization's delivery specifications. An example is the DPP,[2] which has created a detailed QC specification for files delivered to the UK's national

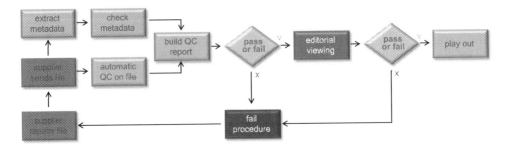

FIGURE 5.5 Example QC Workflow

broadcasters. The other element that is key is to use a common reporting format, so that results of QC processes can be compared and used to automatically trigger appropriate follow-up actions.

Standardizing QC

To make sure QC checks and their results can be compared between organizations, devices, and QC profiles[3] (which may use different parameter settings for the same checks), the EBU has created a large open catalogue[4] of hundreds of QC definitions for the industry to implement in tools and for content parties to create delivery specifications with. Each of the checks is illustrated by a card, which summarizes the key characteristics of what the check is about. A public API allows QC product creators, QC service providers, and users to download the latest definitions.

In fact, in BXF 6.0, a fully standardized XML structure that is completely compatible with the EBU set of QC definitions was added in XML form (Figure 5.6).

Knowledge Base

The QC definitions not only provide detailed technical information, they also include descriptions and references, which help media staff better understand why a check is performed and what specifications it is based on. Examples and test material can be linked to the definitions, as well. The EBU QC catalogue is continuously being reviewed and updated to reflect new content formats such as IMF and platform requirements such as captioning.

QC Outlook

The biggest gain in the short term may lay in the widespread use of standardized QC profiles and QC reports, enabling more integrated workflows. This is reflected in the three main concepts in the EBU.IO/QC data model[5]: QC Catalogue, QC Profile, and QC Report.

We can expect the automation of QC to continue to make use of the latest innovations. Cloud-based QC is already a reality and Artificial Intelligence and Machine Learning promise

FIGURE 5.6 EBU QC Cards

more automation for audio/visual (baseband) checks. The dream of fully automated QC is unlikely to be realized any time soon, though, as especially the visual (baseband) validation of images still is a 'tough cookie' for machines to crack.

Annex A: Classification of QC Checks

A.1 QC Categories

QC checks can be classified in terms of how automatable or reproducible they are. This is especially important when using different QC devices. The following five definitions are used by the EBU.

REGULATORY

A check that must be performed due to requirements set by a regulator or government. Has a published reference document or standard?

ABSOLUTE

A check defined in a standards document including a standard scale. May have a defined pass/fail threshold. As a user, I ought to be able to compare the results of different QC devices and see the same result.

OBJECTIVE

A check that is measurable in a quantitative way against a defined scale, but no standard threshold is agreed. There may be no formal spec or standard, but an agreed method is implied by or exists in the check definition. As a user, I ought to be able to compare the results of different QC devices if I configure the same thresholds.

SUBJECTIVE

The check may be measurable, but with only a medium degree of confidence. A failure threshold is not defined or is very vague. May require human interpretation of results. As a user, I cannot expect different QC devices to report identical results.

HUMAN REVIEW ONLY

A check that can only be carried out by human eyes or ears (Golden Eyes and Golden Ears) or where a human is required to fully interpret the results of an automated QC tool.

A.2 QC Layers

Another way to classify QC checks is the level at which the check is performed. The following definitions are used by the EBU.

WRAPPER

Tests the structure/integrity of the file wrapper, or of the metadata within the wrapper.

BIT STREAM

Tests the structure/integrity of the encoded bit stream, or of the metadata within the bit stream.

BASEBAND

A test applied to the decoded essence – the video frames or audio samples.

CROSS-CHECK

A verification that the values in other layers agree. For example, if a baseband test of video frame size has been completed, and the video frame size metadata in the format and wrapper have been examined, the cross-check will verify that the values match.

Source: https://ebu.io/help/qc/categories

REPURPOSING CONTENT

Monetization is a buzz word in the industry today, tossed about in meetings and forums as a new panacea for ailing media businesses or a grand, perhaps even hidden opportunity for content owners holding assets squirreled away in some dimly lit and cobwebbed library. The idea is that content, with rights owned or acquired, can be repurposed and sold to different markets, different demographics, and across different platforms, maximizing the income from the asset. This sounds like a clever idea, and it can be with forethought and planning, or it can be labor intensive, expensive, and frustrating.

Better than any group in the media and entertainment industry, post-production companies understand this monetization opportunity and its pitfalls and continue to adopt systems to minimize the labor requirements of repurposing media. Here are a few facts to consider:

- Films and television programs are produced in a multiple of technical formats, including standard definition, high definition, 16×9 wide screen, Ultra HD (which is 4K or 8K screen line counts);
- For every film, there are as many as 30 versions edited for content distribution in different parts of the globe where different restrictions apply;
- There are 235+ Video-on-Demand formats used in the United States, and hundreds more used around the world;
- Video codec formats and media file "wrappers" abound and have hundreds of combinations in use in the industry;
- Audio formats are also complicated, with stereo pairs, secondary audio pairs (the Broadcast SAP channels), Dolby 5.1 and 7.1 surround sound, and Dolby Atmos multi-channel sound; and
- Exciting new display innovations such as high dynamic range monitors have added new media packaging requirements to the existing processes.

Dealing with this proliferation of formats, version editing and various types of media is more than full-time work for many post houses. Think about the media on a typical BLU-RAY disk: It usually contains the program video, various audio channels, subtitle selections by language, and the program promotional trailer, and it may be extended with outtake compilation programs and descriptive audio channels, even electronic games and website links to enhance the viewing experience and fan interaction.

We are all familiar with re-purposed media: airline versions of popular, current films; short promotional clips posted to social media sites; video-on-demand; and OTT services like Netflix, Amazon Prime Video, and YouTube. We watch international films with English subtitles or dubbed language versions of Academy Award winning productions. But what does repurposing have to do with our media workflows? Let's examine some of the more important repurposing chores that are executed every day on media assets.

In our commitment to the less fortunate among us, media is consistently re-purposed for accessibility, in particular adding closed captioning for the hearing impaired and deaf, and the relatively new descriptive video audio tracks that provide enhanced storytelling for the blind. In both of these cases, this media enhancement is often mandated by government laws or regulations.

Closed captioning, abbreviated CC, is the set of processes to display text on a television, video monitor, or other visual display to provide interpretive information. The term "closed" indicates that the captions are not displayed on the screen unless viewer selected by the remote-control command or a software menu option, while "open" or "hard-coded" captions are visible to all viewers and are permanently inserted into the video media. Usually, the captions consist of a transcription of the audio channel of the program as it is played or broadcast, sometimes including descriptions of unspoken picture elements. Most of the world does not distinguish captions from subtitles; however, in North America these terms do have different meanings. "Subtitles" assume the viewer can hear but cannot understand the language or accent, or the speech is not entirely clear, so they transcribe only dialogue and some on-screen text. "Captions" attempt to describe to the deaf all significant audio in the asset including spoken dialogue and non-speech information such as the identity of speakers and music or sound effects using words or symbols. Subtitles can be a similar displayed text, the difference between the two being in technical location of the data in the digital or transmitted media. Subtitle applications include providing a written alternate language translation of the main channel audio and this text is often "burned-in" (or "open") on the video and cannot be removed during the viewing process. HTML5 defines subtitles as a "transcription or translation of the dialogue... when sound is available but not understood" by the viewer (for example, dialogue in a foreign language) and captions as a "transcription or translation of the dialogue, sound effects, relevant musical cues, and other relevant audio information... when sound is unavailable or not clearly audible," for example, when audio is muted.

Closed Captioning has come to be used to refer to the North American EIA-608 encoding that is required by NTSC-compatible standard definition video broadcasts which requires that the text be placed in line 21 of the vertical blanking interval in the picture composition. For the US digital television specification ATSC programming, three text streams are encoded in the video: two are backward compatible "line 21" captions, and the third is a set of up to 63 additional caption streams encoded in EIA-708 format. International specifications are considered "subtitles" and their presence can be indicated by the teletext channel reference of "Subtitle 888" on the screen or simply as "888." These notations refer to the PAL-Video standard definition specifications that employed a Ceefax teletext system.

Captioning is modulated and stored differently in PAL and SECAM 625 "line 25" frame countries, where teletext is used rather than in EIA-608, but the methods of preparation and the line 21 field used are similar.

A series of congressional laws require Closed Captioning in the US: The Congressional Act entitled the Television Decoder Circuitry Act of 1990 gave the Federal Communications Commission (FCC) power to enact rules on the implementation of Closed Captioning and required all television receivers with screens of 13 inches or greater, either sold or manufactured, to have the ability to display closed captioning by July 1, 1993. In the same year, the Americans with Disabilities Act (ADA) in which the section Title III required that public facilities, including restaurants, bars, hospitals, and any public location displaying video screens (but not cinemas), provide access to written transcribed information on televisions, films, or slide shows. The Telecommunications Act of 1996 expanded the Decoder Circuity Act requirements to digital television receivers sold after July 1, 2002. Additionally, all Television stations and programming distributors in the U.S. were required to provide closed captions for Spanish-language video programming as of January 2010.

The H.R. 3101 bill, the Twenty-First Century Communications and Video Accessibility Act of 2010, was passed by the United States House of Representatives in July 2010, leading to a similar bill with the same name, S. 3304, to be passed by the United States Senate on August 5, 2010, by the House of Representatives on September 28, 2010, and was signed by President Obama on October 8, 2010. The Act requires, in part, for ATSC Digital decoding set-top box remote controls to have a button to manage the closed captioning display on the screen. This law also required broadcasters to provide captioning for television programs redistributed on the Internet.

Today captions for live programs are transcribed by phonetics-to-text software tools, and the process typically runs a few seconds behind the actual audio channel feed. Live captioning can be considered a form of real-time text, an IP specification. The NORPAK typography extension has overcome many of the initial shortcomings of closed captioning systems and the extended character set now supports all Western European characters as well as Korean and Japanese requirements. The full EIA-708 standard for digital television has worldwide character set support, but there has been little use of it due to EBU Teletext dominating DVB countries, which has its own extended character sets.

When the transcript is available beforehand, the captions are simply displayed during the program after edit work is completed. For prerecorded programs, commercials, and home videos, audio is transcribed and captions are prepared, positioned, and timed in advance. These are the media workflows that most concern our assets. Captions are edited for ease of reading, words substitutions, and reduction of the total amount of text to be presented per minute. Some words can be censored for audiences or to meet broadcast requirements. Video can be edited or "trimmed" to eliminate offensive words or better align with the caption timing.

In an effort to assist global organizations in setting an interoperable standard for captions and subtitles, the Society of Motion Picture and Television Engineers (SMPTE) defined the SMPTE profile of Timed Text Markup Language (TTML), designated SMPTE-TT, to be used for the representation of both "C" captions or "S" subtitles. This specification identifies the features required for interoperability between display systems and to preserve certain semantic features of their original formats, SMPTE-TT also defines some standard metadata terms and some extension types not found in in the basic TTML. The following excerpt from the SMPTE TTML publication provides an overview for our purposes of discussing workflows.

"Known officially as SMPTE ST 2052-1:2010, or SMPTE 2052, the standard provides a common set of instructions for both authoring and distributing captions for broadband video content -- enabling broadcasters to reuse existing TV captions which, in turn, allows them to migrate programming with captions to the Web more easily and duplicate online the experience consumers enjoy on TV today. Labeled the Time Text Format (SMPTE-TT), SMPTE 2052 defines the SMPTE profile of the W3C Timed Text Markup Language (TTML)....regulatory bodies and administrations around the world have, in both technical specification and quantitative requirement, made various forms of timed text (such as captions or subtitles) a necessary component of content. This Standard provides a framework for timed text to be supported for content delivered via broadband means, taking into consideration the economic and practical necessity of the ability to use source data delivered in pre - existing and regionally - specific formats (such as CEA -708, CEA -608, DVB Subtitles, and WST (World System Teletext)) as well as text that may be authored or provided in other formats. It is a goal of this Standard to make timed text essence more useful for broadband content, with equal status to the associated video and audio essence. Considering however that contemporary broadband content on many platforms (e.g. media players, media extenders, home servers, home networks, etc.) may be intended for display in a consumer electronic environment, and that there will be a transition period where such equipment can only utilize pre-existing formats, this Standard also defines functionality for the preservation of legacy formats. It is the intent, however, that new display devices will utilize the more capable TTML timed text structures rather than legacy data formats."

Captioning and the necessary preparation workflows exist for other media including Blu-ray disks and DVDs, cinema presentations of films, and sporting venue live event enhancements.

Generic Descriptive Video Service (DVS) was proposed as a regulation by the FCC but a court ruled that requiring this service by broadcasters was overstepping their authority. Nevertheless, many network broadcasters in the top-25 markets complied with the initial FCC ruling and implemented the service in the secondary audio program, the SAP channel for broadcast programming. The process makes visual media, such as television programs, feature films, and home videos, more accessible to people who are blind or otherwise visually impaired. The term DVS is often used to describe the product itself.

The workflow for assets is relatively straight forward – DVS describers review a program and develop a script that describes all visual elements which are important to understanding the video action and the entire plot. The descriptions and their placement are limited by the natural pauses in the dialogue or are arranged in the time sequence of the accompanying audio to coincide with a particular event or sound effect. The describer then records the new DVS audio channel, and the producer mixes the new recording with a copy of the original soundtrack. This new DVS audio is synchronized with the original video master on a separate audio track for broadcast on the SAP or to its own "DVS master" for home video. For feature films, the descriptions are not mixed with the soundtrack, but kept separate as part of a DTS digital cinema soundtrack.

Multiple Language support: In the United States, it is common knowledge that we support English as a primary language and Spanish as a secondary language, with Federal government enforcement of SAP channel over the air broadcasts. Obviously in our cultural melting pot, this does not serve people of every national origin, and it may not surprise you to know that many specialty language networks exist in North America to address niche populations, including French, Portuguese, Italian, German, Russian, Japanese, Korean, Chinese, and many more. We even have a non-profit network called SCOLA dedicated to re-broadcast of international news programs in their original language for educational and information purposes, covering a wide variety of the world's tongues. Other countries face daunting tasks when dealing with multi-language support. There are 700 individual languages still spoken in the Indian subcontinent and their broadcasters, for example the Times of India networks and Doordarshan (the Indian Government public broadcaster), translate and support broadcasts in 33 languages in an attempt to reach the maximum audience. Typically, Hollywood-based media producers will require their media to be translated into 65 foreign languages for international distribution.

With all these different language requirements, you wouldn't be surprised to know there are post-production companies dedicated to multi-language support, and the workflows required can be quite complex. The primary workflow for assets is either the sharing of a transcript of program's script, or the creation and sharing of a low-resolution media proxy file to enable audio translations of the main dialogue for voice actors to re-create the audio program. Professional voice actors and Foley operators, the people who create sound effects, can be employed for weeks on these specialized projects. New language recordings are sent back to the post-production facility for re-mixing into the original channels to create the new language version, or in some cases used as a "voice over" channel in a new language. Often, these audio workflows require video media editing, to match new audio timing, or to add post program credits (called a post-roll) to list the voice actors.

Similar translation workflows are used to create multi-language subtitle support, although some of the transcription work can be automated by using software to convert phonetics to text and translation software to put the text in different languages. Additional workflows in multi-language versioning are supported by the largest media producers and include translating the on-screen graphics to culturally appropriate referential material, especially when humor requires the audience to be able to read the text in the visual media. This workflow requires video and graphics editing steps, as well as approvals from region-specific experts who can adequately review and assess the cultural impact of the changes in the video media. If you expect your media to be funny in Montreal, you had best have a Quebecois review and approve your new version.

This highlights an important requirement of preparing assets for increased monetization and distribution, the importance of cultural and regulatory considerations. Not all media as originally supplied is welcome around the world. Even within the Unites States there are regional codes and accepted norms which a media executive, in maximizing their use of programs, must support. Most media are put through workflows that address these cultural and regulatory concerns before distribution and these workflows are usually manual operations that address edits and annotations. Media versions can include edits to audio channels to remove offensive language, and edits to video to remove nudity, sexual acts, depictions of violence, or culturally insensitive references or materials. Sometimes on-screen depictions where no rights are held for individuals require the media to mask or hide individuals inadvertently captured in the filming process or people who are in a protected state such as minor children. These workflows can also note and remove product placements that cannot be shared across markets or remove and replace music that is not licensed for a particular region. Primarily these editing workflows are reserved for sensitive content reduction.

Regulations around the world require the display of ratings. While these ratings and parental controls vary from region to region, it is illustrative to consider the methods adopted by the United States to gain an understanding of the requirements and application. In the

Telecommunications Act of 1996, Congress asked the entertainment industry to voluntarily establish a television rating system to provide parents with information about material in programming that would work in conjunction with the new "V-Chip." The V-Chip is a device built into television sets that enables parents to block programming they personally determine to be inappropriate for their families. By February 1996, segments of the entertainment industry joined together and pledged to create a system to support the Government's regulation. They agreed that the guidelines would be applied by both over the air broadcast and cable networks to handle the large amount of programming that would be required to be reviewed, approximately 2,000 hours per day. The Industry advocacy groups, including the National Association of Broadcasters (NAB), the National Cable & Telecommunications Association (NCTA) and the Motion Picture Association of America (MPAA) submitted a revised ratings system to the FCC for review. Under this proposed rating system, television programming would continue to fall into one of the six ratings categories (TV-Y, TV-Y7, TV-G, TV-PG, TV-14, TV-MA). Additional content descriptors of D (suggestive dialogue), L (language), S (sexual content), V (violence), and FV (fantasy violence – exclusively for the TV-Y7 category) would be added to the ratings where appropriate. These ratings icons and associated content symbols would appear for 15 seconds at the beginning of all rated programming. On March 12, 1998, the FCC found that the Industry Video Programming Rating System was acceptable and subsequently adopted the technical requirements for the V-Chip. Similar ratings methodologies have been voluntarily adopted in Australia, Canada, and Europe.

The ratings are stored as metadata associated with the media, and inserted into the media in different locations depending on the dissemination format of the material, digital or standard definition broadcasts, Video-on-Demand files, etc. The workflows that address the ratings must place the data in its proper location embedded in the transmission program to trigger V-Chip activation according to the local regional requirements, or in XML packages that describe and annotate the media file to drive VOD devices or on-screen rating displays of modern digital monitors.

From these descriptions, it becomes apparent that there are lots of workflows that simply transform assets to prepare them for multiple platform distribution. A number of companies founded divisions dedicated to multiple platform support, specializing on creating distribution version of media. Initially this was a labor-intensive process, but it has become increasingly more automated through software innovations. Why are there so many versions required? It is a function of the way the industry grew and changed over time. While the world followed a similar path, it is easiest to use the United States to describe the history of how we got to where we are today, and the workflows that impact media assets because of this historical path.

Community access television began as a way to pool neighborhood resources, share costs, and create a multi-channel antenna system to give homeowners more viewing choices. In the 1950s these systems used microwave antennas to extend signals from distant locations and by the 1960s these systems added C-Band satellite antennas to capture and transmit network feeds. Along the way, entrepreneurs discovered that these community antenna systems were valued by more and more consumers and they began to expand their systems using the local telephone poles to carry the antenna leads – these became the amplified signals we call Cable-TV today. Local municipalities sought to protect their community and offered franchises to the entrepreneurs to maintain a commitment to a minimum level of service. According to the FCC's introduction to Cable Television Regulation,

The Federal Communications Commission first established rules in 1965 for cable systems which received signals by microwave antennas. In 1966, the Commission established rules for all cable systems (whether or not served by microwave). The Supreme Court affirmed the Commission's jurisdiction over cable in United States v. Southwestern Cable Co., 392 U.S. 157 (1968). The Court ruled that "the Commission has reasonably concluded that regulatory authority over CATV is imperative if it is to perform with appropriate effectiveness certain of its responsibilities".

The Court found the Commission needed authority over cable systems to assure the preservation of local broadcast service and to affect an equitable distribution of broadcast services among the various regions of the country.

In 1972, new rules regarding cable television became effective. These rules required cable television operators to obtain a certificate of compliance from the Commission prior to operating a cable television system or adding a television broadcast signal. The rules applicable to cable operators fell into several broad subject areas – franchise standards, signal carriage, network program non-duplication and syndicated program exclusivity, non-broadcast or cable-casting services, cross-ownership, equal employment opportunity, and technical standards. Cable television operators who originated programming were subject to equal time, sponsorship identification and other provisions similar to rules applicable to broadcasters. Like broadcast Television operators, Cable operators were required to maintain certain records and to file annual reports with the Commission concerning general statistics, employment, and finances. In succeeding years, the Commission modified or eliminated many of the rules. Among the more significant actions, the Commission deleted most of the franchise standards in 1977, substituted a registration process for the certificate of compliance application process in 1978, and eliminated the distant signal carriage restrictions and syndicated program exclusivity rules in 1980.

The upshot of this regulatory process was that thousands of small franchised systems were built across the nation, and while major consolidation efforts brought hundreds of systems together under Multiple System Operators (MSAs) control, all of these small CATV systems built their own video distribution plants, called head-ends, each with different vendors and technology for managing clients, set-top boxes, and on-demand programming. Despite efforts by the industry standards organizations like the Society of Cable Television Engineers (SCTE), a plethora of video servers, control tools, satellite integrated receiver decoders, security and permission software systems, display systems, and network-specific products were introduced into these head-ends, each requiring different format support. To get the most value from distributing media, owners needed to create different versions for each platform they wanted to serve. In the US alone, there are over 235 platform-specific required versions for Video-on-Demand files.

Today this distribution versioning is further complicated by the OTT services like YOU-TUBE, Netflix, by Ku-band direct to the home satellite services and their on-demand format requirements, streaming services, and mobile device services. For asset workflows, this versioning chore can be daunting, and many media owners are welcoming the innovations in software tools that support automating these processes. The steps required include formatting the video with transcode systems to support various video system playback requirements, and in some cases actual editing of the video to supply different versions (e.g., offering a PG version of a film edited from an R-rated original). By adding international versioning, a media distributor can manage up to a thousand different media platform formats to support maximizing their income from a program. Each platform format can have complicated workflows associated with its fulfillment.

This is not a new problem of our industry – our standards organizations have been watching the increasing complexity of version requirements with growing dismay for years.

THE INTEROPERABLE MASTER FORMAT

Acknowledging that all the numerous media formats have made it increasingly difficult to distribute and monetize assets, the international standards organizations have sought for decades to design a file format that would support all business models and quality expectations in our industry. Based on constraining the widely used SMPTE standard Master eXchange Format (MXF) and using the lessons learned through the development of the Digital Cinema Package (DCP) used as interchange between distributors and exhibitors for theatrical releases, the SMPTE Interoperable Master Format (IMF) is an international standard for file-based exchange of multi-version, finished audio-visual works. It supports multi-language, multi-reel, subtitles/closed captions, video inserts, and after-the-fact delivery of content with "supplemental packages."

The key concepts underlying IMF include:

- Facilitating an internal or business-to-business relationship: IMF media is not intended to be delivered to consumers;
- While IMF is intended to be a specification for the Distribution Service Master, it can be used as an archival master;
- Support for audio and video, and all media data essence in the form of subtitles, captions, etc.;
- Support for descriptive and dynamic metadata synchronized to an essence;
- Encapsulating (wrapping) of media essence, data essence as well as dynamic metadata into well-understood temporal units, called track files using the MXF (Material eXchange Format) file specification; and
- Each content version is embodied in a "Composition," which combines metadata and essences. An example of a composition might be a theatrical cut or a VOD edit.

IMF is built upon a core framework (SMPTE ST 2067-2) that includes essence containers, audio and timed text essence formats, basic descriptive metadata, complex playlists, delivery, etc. According to the SMPTE introduction to the specification, "This core framework is supplemented by incremental constraints ("applications") specific to particular domains. A typical application will specify video codecs and image characteristics, and can specify additional descriptive metadata, for instance. By sharing the core framework across applications, IMF can adapt to evolving industry requirements while maintaining more than 95% commonality across implementations and practices."

SMPTE has spearheaded the development of different specifications for different use, calling them applications. IMF Application 2 (SMPTE ST 2067-20) supports High Definition (1920×1080 frame size) Standard Dynamic Range (SDR) images and JPEG 2000 image coding. IMF Application 2e (SMPTE ST 2067-21) extends Application 2 with support for High Dynamic Range (HDR) images up to 4K frame size. Applications 2 and 2e have received support from studios, post-production facilities as well as the manufacturing community, and is being used as a delivery format to OTT services (Netflix). Application 2 and 2e are often referred to collectively as IMF Studio Profile, and the SMPTE sponsored "plug fests" where vendors test the interoperability of systems designed for compliancy are focused on the IMF Studio Profile. Other Applications exist, such as Application 3 (SMPTE ST 2067-30), while others still are being developed. In July of 2017, SMPTE and the UK-based Digital Production Partnership (DPP) agreed to partner to develop an IMF application specification for EU Broadcast and Online services with an eye to publish a specification in 2018. IMF is designed to support all possible international distribution versions (special editions, airline edits, alternative languages, etc.) of a high-quality finished work, including features, trailers, episodes, advertisements, etc. International media distributors such as Netflix and Sony Pictures Entertainment have standardized on the IMF for receipt and distribution of content. To quote Bruce Devlin, SMPTE Governor for the U.K. Region, "These features and the specification which is built upon the IMF standard overall are critical to the realization of file-based interoperability on a large scale, as they ensure that broadcasters can use IMF workflows with their existing content archives."

Behind the technical specifications are solid methodologies that organize the media. When designing this specification, the SMPTE IMF committee went further than the original MXF group – it defined IMF as a new approach to media file organization and management, in other words a new way to think about the files that support better interoperability and version distribution. There are three key concepts that IMF introduces: Component Play Lists (CPLs), Output Play Lists (OPLs), and Identifiers.

A master version is defined in a Composition Playlist (CPL) which summarizes the playback timeline for the asset composition and includes metadata applicable to the composition in its entirety via XML. It is important to note that the IMF specification allows for the creation of many different distribution formats or versions from the same composition. This can be accomplished by specifying the processing/transcoding instructions through an Output Profile List (OPL). The OPL is a set of actions – the transformations that a CPL requires for a particular version, including format conversions, digital rights management, content delivery network preparations, etc. One CPL with an OPL defines an IMF complaint version of the asset.

The CPL is not designed to contain essence but rather to reference external Track Files that contain the actual essence. This construct allows multiple compositions to be managed and processed *without duplicating common essence files* – this feature of the methodology is important as it leads to efficiencies and cost savings which are illustrated later in this section. The IMF CPL is constrained to contain exactly one video track (image sequence). This flexible CPL mechanism decouples the playback timeline from the underlying track files, allowing for incremental updates to the timeline when necessary, to change an audio or subtitle track without impact to the entire package.

Each CPL is assigned a universally unique identifier (UUID) that tracks versioning of the playback timeline. Resources within the CPL reference essence data via each track file's UUID.

A truly important concept introduced by the IMF committee is the idea of "Identifiers." IMF media is comprised of many identifiers, on many different levels, and they are used to build a particular composition for a CPL, or mark segments of media or specific events on a timeline in the media. These identifiers can act like pointers, and an IMF compliant file can identify multiple sources of media to be "stitched" together to form a new segment for a distribution requirement. Simple content editing, splicing in advertisements or adding pre- or post-rolls is a straight-forward assembly chore in an IMF compliant media preparation system. These identifiers provide tracking mechanisms on many layers, and tracking tools enable workflow automation, making the process of building interoperable versions automatic with all the benefits computerization offers, making operations faster, easier, more efficient, and less labor intensive. This use of identifiers also enables new processes, such as sending identified linked components to change just a part of a previously delivered IMF file (a "supplement").

By design, asset delivery and playback timeline aspects are decoupled in IMF. The delivery entity between two businesses is called an Interoperable Master Package (IMP): IMF media is wrapped for distribution, and the specification details the specific requirements of that package.

An IMP is described as including:

- One Packing List (PKL – an XML file that describes a list of files), and all the files it references;
- One or more complete or incomplete Compositions

  - A complete IMP contains the full set of assets comprising one or more Compositions
  - A partial IMP contains one or more incomplete Compositions. In other words, supplemental media not referenced by the PKL

Implications of IMF for Asset Management and Workflows

To maximize these advantages, Media Asset Management systems must be able to offer an end-to-end IMF workflow that leverages the IMF specifications without overloading processes with internal conversions or limitations in the content management structure. New version generation needs to be managed efficiently in order that no media is replicated until required: different versions should merely reference the media masters in unique compositions that includes the additional elements needed for new material; i.e., a localized version in French with text insertions and audio and subtitles in French; another commercial version with different edits; a different technical version; etc. And finally, content delivery must fit seamlessly into IMF specification compliancy so that the package that is delivered takes the advantages of all these "Logical Versions," supporting a flexible mechanism to select the proper components to deliver (video, audio, subtitles), in the correct order and with the transformations to be accomplished for a particular distribution reception site.

According to Julian Fernandez-Campon, the Chief Technical Officer of MAM Solution vendor Tedial, "To summarize, new Media Asset Management Systems need to be able to support:

- IMF Import, processing the IMF package and creating assets that can be physical or virtual, preserving and extending the IMF Composition playlist without media duplication
- Content Enrichment, by creating new versions establishing logical references to the Master and taking advantage of the logical references to the media from the Master(s)
- IMF Delivery not limited by the original IMF received packages, ready to generate new compliant packages with any compatible combination of the components for a specific Title: Video (SDR, HRD, different aspect ratios), Audio Languages, Subtitles.

- IMF end-to-end workflows imply the full support of the IMF packages at all levels, not 'IMF-ish' MAM that stores and manages the media using IMF labels without actually processing the assets and maximizing the advantages of content management and exchange.

An end-to-end IMF workflow requires these stages:

- IMF Import: IMF Content is received as a complete unit, or with additional supplemental packages, and is ingested into the MAM;
- Content Enrichment: full support for the creation of new logical versions or new component aggregation, such as adding subtitles or audio tracks that were not received as IMF packages; and
- IMF Delivery: Content can be delivered as the original IMF package received, or as new IMF packages based on logical combinations of the components, supplements and assemblies."

For a successful IMF compliant MAM deployment, content enrichment must be considered a fundamental part of the workflows. Content enrichment includes all the processes and asset workflows that occur inside a media operational facility to enhance the received content and increase its value. In practice, there is limited content exchange based on an originally distributed IMF package, as every media company and broadcaster have their unique methods of enhancing the value of their incoming media. While there are many types of content-enrichment activities and an IMF compliant MAM supports them all, the most common scenarios include subtitles aggregation, audio tracks aggregation, and new versions of the image sequence.

Subtitles aggregation occurs when subtitles are generated externally to be attached to a particular title in the MAM. Subtitles are received as files in a specific format, converted if needed into IMSC1 to comply with the IMF-constrained specifications, and attached to the Asset Title within the MAM library hierarchy. If subtitles aggregation is required, it's crucial that the MAM system can expand its asset management to display and manage all the multiple subtitle languages and types to be supported.

Audio tracks aggregation as individual files is similar to the subtitles case with the difference that there are more variants due to the sonic layout (2.0, 5.1, etc.) and the class (Full Mix, Music and Effects, etc.). In the IMF compliant MAM, audio tracks must be attached to the Asset Title and a proxy generated.

New Versions can be created within customer facilities using MAM tools to generate new localized versions (trimming, adding text insertions, etc.) that are related to the main title with a specific, annotated relationship. New version workflows present several challenges, depending on the customer business rules, such as slate insertion and any formatting of the video content including bars, audio tones, blacks, etc., that might need to be done with or without an off-line editor. A key feature for an IMF compliant MAM is to manage new version assembly or control the import of the EDL from an editor to identify new fragments and create a logical asset referencing the Master in the MAM. IMF identifiers can be crucial to streamlining and automating this operation.

IMF Publishing is the other key aspect of an end-to-end IMF workflow. To take a very simplistic view, the MAM system needs to be able to "export" IMF packages. One simple and straightforward approach is to "route" the imported IMF packages through the MAM and deliver them in the same compliant package as they were received, yet this simple workflow is not enough for a true end-to-end IMF workflow as the following features are in demand by media companies:

- Export both Complete and Supplemental IMF Packages: To take the most advantage of the IMF capabilities and cost efficiencies, masters will be delivered as a complete IMF package and the localized or editorial versions as supplemental. This decision depends on customer business rules and their recipients' requirements – in some cases a reception site requirement might force the export of a localized version as a complete package. Usually, a mix of complete and supplemental packaging is required;
- Delivery of new combinations of audio/subtitles: The media company may need to deliver a combination of audio/subtitles that was never received in the original master IMF package, e.g., French version with English subtitles. Typically to keep their costs low, a translation company only delivers a French supplemental file that references the original master;
- New IMF package generation, for those versions generated in-house that did not arrive as IMF Packages;
- New components export in the IMF Package to allow the generation of new packages with a combination that was never received (e.g., English 2.0 + English 5.1 + French 2.0 + French 5.1); and
- Transformations: Reception site–specific requirements for generation of site unique audio routing, video transformations, etc., based on the OPL delivery profiles stored in the IMF compliant MAM.

Customer Business Rules outside of IMF: IMF is standardized, well defined and is great for content distribution. The specification permits some level of freedom for customer-specific applications according to their needs, which implies some business rules must be designed to be compliant with the IMF requirements. When implementing workflows, the challenge for a MAM system is the solution must continue to allow business flexibility at the same time the core IMF support is maintained. Some examples that are in use in media workflows include custom metadata fields that become data model extensions; the co-relation of multiple masters (SDR, HDR, 16:9) that arrive as independent IMF Complete packages, yet need to be related in the MAM; restrictions in formats, for example where the customer does not allow HDR versions for specific media; specific essence conventions in the video/audio, to insert slates, tones, blacks, etc.; and associated media such as artwork, photography, trailers, and other related components that need to be linked with the title. All these business rules are outside the native IMF support in a compliant MAM. A successful solution can embed these variations within the workflows to comply with particular media companies' business requirements and offer an efficient IMF end-to-end workflow tailored to specific needs.

The Business Case for IMF

Examples of the application of IMF principles can be drawn from multiple sources, such as this case study from the Netflix techblogentitled "House of Cards Season 3:"

> Netflix started ingesting Interoperable Master Packages in 2014, when we started receiving Breaking Bad 4K masters. Initial support was limited to complete IMPs (as defined above), with constrained CPLs that only referenced one ImageSequence and

up to two AudioSequences, each contained in its own track file. CPLs referencing multiple track files, with timeline offsets, were not supported, so these early IMPs are very similar to a traditional muxed audio / video file.

In February of 2015, shortly before the House of Cards Season 3 release date, the Netflix ident (the animated Netflix logo that precedes and follows a Netflix Original) was given the gift of sound.

Unfortunately, all episodes of House of Cards had already been mastered and ingested with the original video-only ident, as had all of the alternative language subtitles and dubbed audio tracks. To this date House of Cards has represented a number of critical milestones for Netflix, and it was important to us to launch season 3 with the new ident. While addressing this problem would have been an expensive, operationally demanding, and very manual process in the pre-IMF days, requiring re-QC of all of our masters and language assets (dubbed audio and subtitles) for all episodes, instead it was a relatively simple exercise in IMF versioning and component-ized delivery.

Rather than requiring an entirely new master package, the addition of ident audio to each episode required only new per-episode CPLs. These new CPLs were identical to the old but referenced a different set of audio track files for the first ~100 frames and the last ~100 frames. Because this did not change the overall duration of the timeline, and did not adjust the timing of any other audio or video resources, there was no danger of other, already encoded, synchronized assets (like dubbed audio or subtitles) falling out-of-sync as a result of the change.

The impact of IMF principles and methodologies can result in significant efficiencies and costs savings. **Francois Abbe**, CEO of Mesclado of France, commissioned multiple case studies to investigate the application of IMF to current workflows and the following summarizes the results of the case study #1 inquiry:

Case Study: Film Mastering and Versioning

SUMMARY: COST SAVINGS 26% USING IMF

Background: IMF was born of the D-Cinema Package (DCP) standard success. IMF enables film distribution worldwide using a single package, each region receiving one or more versions of the content, also known as "Compositions." Several compositions can share common content, thus reducing distribution complexity. In 2006, the Hollywood studios began to consider a similar format for their masters.

The Challenges encountered include reducing interoperability issues, supporting the latest video formats like UHD, HDR, and VR, and enhancing multi-versioning support whilst keeping infrastructure costs down. Outcomes of the study included automated versioning and delivery based on metadata, the ability to set archiving masters, and producing outputs in multiple delivery formats.

Our inquiry approach was to compare the operational expenditures from mastering to versioning, both with and without IMF in the following areas:

1   Production

2   Mastering

3 Versioning
- Dubbing & Subtitles
- Trimming & Credits

4 Storage

5 Distribution

Production media:

- One-hour documentary
- Two versions: Original Version (OV) and International Version (IV)

Mastering:
Without IMF

- Using ProRes 422 to produce one MXF or Quicktime file per version
- File size: 60GB per version

With IMF

- Using JPEG-2000 to produce the Interoperable Master Package (IMP) with higher quality images
- File size: 100GB for the package

Versioning:

- For versioning purposes, a proxy is created. The proxy is sent to a lab along with a subtitling or dubbing order.
- Dubbing and subtitling: We order five subtitles and five extra audio languages (English, German, Italian, Spanish, and Arabic).
- Without IMF and With IMF, in both cases we receive five .stl files and five audio files
    - Five new masters are produced for the dubbed versions (60GB each)
    - Supplemental packages are added to the original IMP with new audio and subtitles (~100GB total)
    - New CPLs are created for each version

Trimming & Credits:

- We create two new versions for Arabic and German territories: first for censorship purposes (cutting scenes) and the second for co-production contract (adding credits)

Without IMF

- Two new masters are produced.
- The old versions are purged.

With IMF

The IMF package is updated:

- German version: new CPL with extra credits (slate)
- Arabic version: new CPL with a removed scene

Storage:

Without IMF

- We store the OV, IV, German and English versions.
- Other language versions are purged to save cost (60GB each)

With IMF

- We store the IMF package. All the versions are included in one single IMP (~100GB total)

Distribution:

Without IMF

- When ordered, each master is used to produce the version delivered to the client
- Spanish and Italian versions are purged after use to make savings on storage

With IMF

- When ordered, the IMP is used to produce the version delivered to the client based on the CPL
- All versions are kept within the IMF package

Three years later an order is received for dubbed Spanish and Italian versions:

Without IMF

- Two new masters are produced for the dubbed versions (60GB each)

With IMF

- The IMP is used to produce the version delivered to the client based on the existing CPL

Analysis and conclusions of the study:

- With all costs computed using a typical post-production rate card, total cost of owner-ship over four years is predicted to be:
  - TCO without IMF = $29,904
  - TCO with IMF = $22,017
- Cost savings:
  - Storage cost is very small compared to the costs of editing and producing new versions (subtitling or dubbing), so it is wise saving all versions

- Storage cost per version per year
- The costs of saving extra versions is marginalized with IMF:
- Cost without IMF = $14.00
- Cost with IMF = $0.40
- Cost savings 26% using IMF over non-IMF methods

AUTOMATING WORKFLOWS

As we have learned in the proceeding sections, media enrichment and versioning maximizes revenue, that "monetization" that each media and broadcasting company seeks to extend their bottom line profitability. Our development of workflows for our assets was predicated upon the business rules that each unique company has evolved to support their exclusive business models. In the beginning, all workflows started with humans doing their jobs manually, moving physical media before the advent of digital tapeless processes, then adapting to more software-based systems and digital media as it became viable in the industry. As the systems we use grew in complexity and our libraries increased in size and scope, automation of the operations became an important business driver for our organizations. Efficiency and cost effectiveness became the evaluation criteria for our workflow designs and today we focus our efforts on removing steps and enabling streamlined functions for human interaction while automating the workflows as much as possible. In a perfect system, everything would be automated. This section explores the underlying requirements for automation, the basics of automating workflows at the machine level and at the human level, monitoring the metrics, and the application of Artificial Intelligence and Machine Learning to enhance our workflow operations.

Like the Master Control automation, workflows automation requires deep integration into both the physical or virtualized IT network infrastructure and all the third-party tools in the proposed workflows or the entire media factory. Third-party tools can include both physical hardware like audio and video routers, media switchers, satellite integrated receiver-decoders, and computer servers, and software-based tools like transcoders, automatic quality control, content delivery networks such as Aspera, Signiant or File Catalyst, watermarking and digital rights management, etc. Integrations to workflows systems enable the workflow management software to control the steps that require external tools and orchestrate the overall stages into a complete process.

In order to organize our discussion of workflow to platform tool integrations, the topics can be separated into a few categories:

- One-way vs. two-way interfaces
- Simple command/control or reporting systems (SNMP traps)
- Defined interface standards like BXF
- APIs
- Web services, SOAP, and REST

One or two-way integrations determine the complexity of the interface, the set of commands supported, and the scope of processes that can be automated. One-way interfaces are usually for devices, many of the older designs still in use, that support commands from an external system. These tend to be a set of commands structured with no answer protocol, a "fire and forget" process. These systems are phasing out as new technology becomes available, or their operation in a particular workflow does not depend on responses to complete. Examples include some content delivery network operations where an asset is simply posted to an FTP site or a watch folder, and the CDN transmits the files according to a pre-set profile for deliveries. Two-way communication between integrated devices is becoming more common, where response and reporting protocol is returned through the interface to update the workflow engine or orchestration tools.

Simple command/control and SNMP were some of the first integrations for workflow and system monitoring. Simple command and control is a one-way integration scheme, for example, commands to control a video tape recorder (VTR) is typically a single direction command set with an interface connect specification (i.e., RS422) that manages a simple set of control commands to play, stop, fast forward, rewind, find/set a tape location, etc. Another example of simple command and control is media switcher control which consists of an interface and commands to route the video and audio within a plant. These one-way systems may provide simple reporting back to the workflow orchestration software. Some systems rely on Simple Network Management Protocol (SNMP), a standardized IP protocol for collecting/organizing information about networked devices, and for modifying that data to change end device operations. Devices that typically support SNMP include cable modems, routers, switches, servers, workstations, printers, etc.

SNMP is widely used for network monitoring. SNMP provides management data in the form of variables on the managed systems organized in a Management Information Base (MIB, a type of database) which typically display the system configuration and current status in a graphical, topographic view. Variables can be remotely queried or "pinged" for a status update and managed by supervision applications. SNMP has been updated over time to add feature improvements in performance, flexibility, and security and version 3 is most current. As a standard for integration, SNMP is a component of the Internet Protocol Suite as defined by the Internet Engineering Task Force (IETF) and it consists of a set of standards for network management, including an application layer protocol, a database schema, and a set of standard data objects.

Integrating traffic systems, program schedulers, Master Control automation, and MAM-workflow tools, the current SMPTE BXF protocols include an XML schema definition (XSD) collection for schedules, as-run, content, content transfers, etc. Connecting the Master Control and traffic departments is the most common broadcast use. When properly implemented, BXF-based applications automate the flow of data between the systems increasing process efficiency, streamlining manual steps, maximizing the value of content inventory, and increasing flexibility for the sales department and client advertisers.

As an XML-based communication schema, the two-way BXF interface allows for near-real-time messaging and updating between disparate systems. The XML-based messages include instructions about program or interstitial changes, allowing an automated approach to as-run reporting and schedule changes in the connected systems. Other BXF capabilities include near-real-time dub orders, missing spots reports and content management.

In computer programming, an application programming interface (API) is a set of subroutine definitions, protocols, and tools for building integrations between specific software modules. In general terms, it is a set of clearly defined methods of communication between software components. A solid, well-defined API makes it easier to develop a computer program by providing building blocks, which are strung together by the programmer. There are many types of APIs, and they may be for a web-based system, operating system, database system, computer hardware, or software library. Because of this diversity, API specifications can be complicated and varied, and often include specifications for routines, data structures, object classes, variables, or remote calls. Documentation for the API is usually provided to facilitate usage.

A web service is a software service offered by one electronic device to another electronic device, communicating with each other via Internet Protocol (IP). It is a software function provided at a network address over the web, with the service always "on." In a web service, the Web technology such as HTTP – originally designed for human-to-machine communication – is utilized for machine-to-machine communication, more specifically for transferring machine-readable file formats such as XML and JSON. In real-world practice, a web service typically provides an object-oriented web-based interface to a database server, to be employed by another web server or by a mobile application, that in turn provides a user interface to an end user. For the purposes of workflow integrations, a web service is a software system designed to support interoperable machine-to-machine interaction over a network.

Web services may use SOAP (Simple Object Access Protocol) over HTTP protocol, allowing less costly (more efficient) interactions over the Internet than via proprietary solutions. Besides SOAP over HTTP, web services can also be implemented on other reliable transport mechanisms like FTP. Developers identify two major classes of web services: REST-compliant web services, in which the primary purpose of the service is to manipulate XML representations of web resources using a uniform set of "stateless" operations; and arbitrary web services, in which the service may expose an arbitrary set of operations. The term "web service" describes a standardized way of integrating web-based applications using the XML, SOAP, WSDL (Web Services Description Language), and UDDI open standards over an Internet Protocol backbone. XML is the data format used to contain the data and provide metadata around it, SOAP is used to transfer the data, WSDL is used for describing the services available, and UDDI lists what services are available.

Many organizations, especially asset media workflows use multiple software systems for management, control, and orchestration. Different software systems often need to exchange data with each other, and a web service is a reliable method of communication that allows two software systems to exchange this data over the Internet. The software system that requests data is called a service requester, whereas the software system that would process the request and provide the data is called a service provider. A directory called UDDI (Universal Description, Discovery, and Integration) defines which software system should be contacted for which type of data. When one software system needs one particular report/data, it goes to the UDDI to discover which system it can contact to request that data. Once the software system confirms which system to contact, the process to contact that system uses a special protocol called SOAP. The service provider system would first validate the data request by referring to the WSDL file, and then process the request and send the data under the SOAP protocol.

Different software may use different programming languages, and hence there is a need for a method of data exchange that doesn't depend upon a particular programming language. Most types of software can, however, interpret XML tags. Thus, web services can use XML files for data exchange. Rules for communication between different systems need to be defined in he published API documentation, such as:

- Parameters for systems to request data from each other
- Specific parameters required in a data request

- The defined structure of the data produced (Normally, data is exchanged in XML files, and the structure of the XML file is validated against an .xsd file.)
- Error messages to display when a certain rule for communication is not observed, to make troubleshooting easier
- All of these rules for communication are defined in a file called WSDL, which has a .wsdl extension

Once integrations between devices can be established, we can map out the processes in our operation and look for points where the step is routine and repeatable as a location for automation. Workflow automations can take many different forms, beginning with the low tech but reliable method of employing Watch Folders and scripts. Pioneered decades ago, but still in use today, watch folders, sometimes called "Hot" folders, are receptacle/storage location constructs that wait for an object/ file or a pointer to an object/file to be placed in the storage location and trigger an action. "Scripts" are small coded action steps to be run against any asset or pointer to an asset that is placed in a particular "folder." Scripts can be simple or complicated. The drawback to this method of triggering workflows is it is fairly rigid, offers little monitoring opportunity, and can be complicated to maintain. While many workflow automation systems use watch folders, primarily for ingesting newly arrived media, the development of more powerful workflow engines has eclipsed the watch folder method of automating operations.

Workflow engines operate at two levels: an abstracted, integration-media processing layer and the Business Process layer. Around the turn of the millennium, workflow software manufacturers began to separate their machine-to-machine management chores from their human processes. This led to innovations in integration and media movement tools. Building a technology stack with an abstracted integration and media processing layer removes the requirement for "hard coding" the integrations directly to the human processes. Previously, systems built without an integration layer were expensive to update or change workflows because the interfaces to third-party tools had to be re-written as part of the workflow orchestration. Separating the integration layer provides more flexibility and allows for automation of media movement behind the scenes, hidden from the average user.

People power: most workflow orchestration tools use Business Process Management (BPM) to map and manage their user, human-driven workflows. BPM employs a standardized design graphic set that makes it easy for all types of users to map their operational steps. The Internet standard Business Process Model and Notation (BPMN) provides businesses with the capability of understanding their internal business procedures in a graphical notation and gives organizations the ability to communicate these procedures in a standard manner. The current version of the specification is BPMN 2.0. The graphical notations facilitate the understanding of the performance collaborations and business transactions between departments and organizations to ensure that businesses have internal understanding between themselves and their participants in their business and enable organizations to adjust to new internal and B2B business circumstances quickly. Business Process Management Initiative (BPMI) developed BPMN, which has been maintained by the Object Management Group since the two organizations merged in 2005. Version 2.0 of BPMN was released in January 2011 at which point the name was adapted to Business Process Model and Notation as execution semantics were also introduced alongside the notational and diagramming elements.

BPM workflows can also be built to include unique business logic to further automate machine-measured decision processing. Third-party tool capacity monitoring and storage retention policies can be automated, and defined, measurable actions can be machine initiated based on criteria established by management. Examples include:

- Sixty days after last usage, move media to tape storage;
- Five days after the project was passed to an editing workstation, send email reminder about project due date to assigned user;
- Two days before a transmission event, alert user that media has not arrived to fulfill placeholder
- Etc.

What's the difference between workflow and orchestration? It's a matter of breadth and scope, as well as a little "marketing speak." Workflows are a series of serial or branching parallel steps depicting an operation, while orchestration is a collection of nested workflows that organize a department's chores, or a complete media factory end-to-end functionality. Every step, in a workflow and in an orchestration, can be managed and potentially automated.

Following the Deming management methods, task assignment to and management of users and user groups allows for closer management of human steps. Setting a task list for the human operators and controlling the steps of their BPM workflow ensures consistency across the staff. Activities can be monitored and adjusted to meet and exceed key performance indicators (KPIs) as well as administer appropriate leadership to personnel though accomplishment analysis.

Automation brings about measurements, reports, and metrics, allowing managers and executives to customize reports to their business goals. Dashboards can be run by external data-mining tools monitoring the database or display reports run every few seconds to update the data. Most orchestration workflow software has built in escalation processes which use business logic or workflow failures to trigger procedures designed to draw attention to the issue, failure, or delay. Email messages, SMS texts, and Dashboard alarms are examples of escalation procedures in workflow systems.

As media companies work to build full orchestrations, they are often faced with deciding which of their business systems will be the financial system of record. Many companies are finding that slaving workflow engines to external systems, such as work order managers, traffic and billing systems, Rights Management systems, or Content Management systems (CMS), lets their leading business financial system act as the dominant system to control the operations and personnel workflows. Tying their business systems to the workflow engines and MAM databases allows for deep data mining and analysis, providing unique insights to the business. Additionally, some companies use customer and client portals with search/retrieve/publish workflows dedicated to the behind the scenes operations, all tied together with a financial system for reporting and billing so they can maximize every opportunity to sell their content and assets.

In 2017 Artificial Intelligence and Machine learning began to be integrated with workflow engines to reduce labor and create more efficient, faster operations. These tools can be linked to social media and ratings systems to provide feedback and better tune their operations. Examples of AI in use in workflows today are:

- Automated metadata annotation on ingested assets
- Acquisition of external metadata to apply to the media (EIDR data, IMDb, etc.)

- Program rating performance influencing live production; and
- Live sports event EDL workflows to automatically capture, clip, and disseminate media to fan platforms and social media.

The key to successful asset media workflows today is automation and orchestration and a consistent plan for continual improvement based on statistic and measurement.

CLOUD IMPACTS

The joke passing around a 2017 SMPTE conference was "the cloud is just someone else's infrastructure." While this was meant as a disparaging comment, it actually highlights a benefit of public and private cloud operations. In a recent IABM conference presentation, the VICELAND network described their real-world experience of implementing from their London headquarters an entirely new network linear play-out channel on the other side of the world, in New Zealand, in only six weeks' time from start of project to live broadcasts. New technologies are disruptive, and the public and private cloud technologies are re-inventing business models worldwide. Our media industry is no different – the Internet and IP infrastructure has evolved to reliably support viable video quality of service requirements for broadcast and media distribution, as well as a host of production services. This is a boon for our media workflows.

While this technological revolution has increased the direct to consumer options and is quickly changing the economics of our industry, many of the benefits of cloud computing are variations of already tried and true methodologies used around the world. For example, European broadcasters have been outsourcing their transmission services to carriers and system providers for years, and this model has proven to be financially viable – cloud infrastructure is an extension of the framework that the service providers already manage with service level commitments to nearly 100% availability. As United States broadcasters and media companies look to better control costs and increase flexibility in their offerings, the cloud holds great promise.

According to Accenture, there are four potential benefits of cloud computing for broadcasters:

1. Faster speed to market, which closes the gap on the faster service delivery cycle of the OTT entrants;

2. Scalability to handle spikes in workload, including live events, and surges in the popularity of new services;

3. The ability to collect, store, and conduct analytics on vast amounts of data, generating insights to drive personalization, service development, customer experience, and one-to-one relationships; and

4. Driving on-going service innovation through agile development, constant iterative experimentation, and a culture of "fail fast and fail cheap – then move on."

Media owners and broadcasters are faced with significant business model changes that drive cloud computing as a necessity: Consumers expect more choices and demand more customization, the modern broadcaster needs to react to these demands quickly and confidently, and the need for scalability of computing power that only a virtualized and extensible computing platform can provide. Fragmented viewing audiences are triggering changes in content preferences and our industry is adapting by embracing flexibility in managing complex business and revenue models to maintain and grow engagement across demographics.

Everybody wants everything yesterday – consumers "binge watch" every episode of a new series and quickly demand another season. The reality of these time pressures underlies the content delivery market today. Media workflows must reduce production and distribution cycles to reduce time to market for new assets and experiences, as well as consistently increase their existing libraries of content.

Cloud computing services are typically a combination of utility computing, storage, and network bandwidth with no regional limitations to implementation, hence the confusing and sometimes misused moniker, "cloud." Because our industry is not new, we are adapting on-premises workflows to the new infrastructure, and the jargon most often used to describe the cloud operations is "Public," "Private," and "Hybrid."

Public cloud infrastructure is typically utility services that are used by businesses to buy computing, storage, and bandwidth on a fixed monthly rate or on-demand. Some more mature cloud services companies have been offering services on-demand, including transcoding, versioning, editing, and distribution workflows. The designation of "Private" cloud refers to corporations that want or need to manage their own data and computing processes in their privately purchased environment, restricting access and maintaining tighter security access. The word "Hybrid" has come to have varying meanings in our industry, depending on a particular individual's design, either a mix of private and public virtualized environments, or a mix of on-premises hardware and software systems, local virtualized infrastructure and public cloud services. One clear fact emerges from the market research on cloud computing in our industry, the most common cloud deployment in media workflows is the hybrid approach.

Content drives our workflows and the type of infrastructure does not change this core fact – the key to a successful move to a cloud-based operation is to ensure that your management team can rely on consistent and predictable financial savings and the security of your media, while taking advantage of the great flexibility and scalability the public virtualized infrastructure. Adoption of cloud has taken two paths: the first is the use of web scale suppliers such as Amazon Web Services, Google Cloud Services, and Microsoft Azure for hosting and delivering media. The second level of adoption is leveraging the cloud stored content for virtualized production, post-production versioning, and distribution workflows. Today cloud clients are adapting their workflows to manage the content flow from creators to distributors and broadcasters, move content to post-production services, post the materials and assets to specialized distribution platforms and, of course, support the growing number of direct to consumer options for distribution.

Start-up companies, like the VICELAND IABM case study referenced at the beginning of this section, are the greatest beneficiaries of a globally coordinated infrastructure leveraging Just-in-Time scale up and down operations. Other large broadcasters and media owners are employing cloud computing in response to market dynamics for "pop-up" channels, new or temporary workflows, and to address customer demand. These large corporations are finding innovative ways to compete with their nimble OTT competitors and use the super-fast cloud deployment speed to maximize their competitive advantages. There is no avoiding the fact that surviving in today's media world requires a global mind-set, and has become an increasingly complex, expensive business. Teams distributed worldwide, thousands of content distribution outlets, diverse technology options, and the ever-increasing fragmentation of viewership are driving tremendous challenges in managing cost-effective global operations.

Broadcasting flexibility and cloud-based linear channel playout are now proven workflows. The public service providers have demonstrated their environments' scalability to quickly adjust virtual infrastructure to the ever-changing requirements for digital content services, often permitting content rights holders the chance to cash in on overnight sensations where popularity for specific content products drives demand. While cost pressures in our market seem to intensify, driven by the entry of agile low price players, increasing rights costs and shortened technology life-cycles making expensive IT hardware obsolete, an interesting development is that "cost" of cloud services has become less important when compared to other benefits, especially the improvement to delivery mechanisms, and offering today's media executives the opportunity to experiment with new ventures without incurring capital expenses and the opportunity to quickly shut down a trail that doesn't meet expectations. Broadcasters and media distributors are using the public cloud to avoid upfront technology investments and align costs with revenues and cash flows. With the uncertainties of today's industry, it's a very appealing management decision to employ a "pay for what you need, when you need it" subscription model that cloud offers.

Big Data and the Internet of Things is now of increased concern for today's modern broadcaster, and with the application of Artificial Intelligence and Machine Learning the storage and analysis of the tidal wave of data will drive cloud applications. More than just program ratings, content personalization, fan engagement, support for niche audiences, enhanced user experiences and the engagement that creates a one-to-one relationship with consumers will better connect our media companies to their revenue source. And most, if not all, of these services will be automated workflows based in cloud infrastructure to take advantage of the massive computing power needed for analysis.

Asset workflows in the cloud bring measurable benefits. The tapeless operations we have been building over the past decade can quickly adapt their workflows and libraries to the cloud infrastructure. Cloud services can lower the on-going cost of these functions. Live event coverage and the cost of remote operations, challenged by increased rights as well as production costs, will drive sports broadcasters to embrace IT resourcing in less investment-intensive cloud computing platforms. Video streaming and on-demand services are maturing, finding more ways to leverage richer content and using cloud workflows to better engage and attract new consumers. And a chief benefit of employing cloud computing is using the homogenizing efforts of the cloud suppliers to take advantage of the increasing compatibility of the platforms, allowing media companies to profit from rich content across all devices, smartphones, tablets, hybrids, and televisions.

NOTES

1. Nielsen's 2014 Advertising & Audiences Report.
2. https://www.digitalproductionpartnership.co.uk/publications/theme/quality-control/.
3. A QC profile is a collection of QC checks (and relevant configuration parameters). Users use such profiles in QC products a bit like a 'macro.' It saves them specifying the same

settings over and over again, and it limits the room for errors. An example is using a specific 'Ingest QC Profile' to verify incoming content complies with the media company's own delivery specification.

4. https://ebu.io/qc (CC-BY 4.0).
5. The original EBU.IO/QC data model was inspired by FIMS QA, and inspired BXF QC.

CHAPTER 6

Distribution to the Viewer

Section Editor: Glenn Reitmeier

The essential step of how to get the content to the viewer is the focus of this section. Distribution systems have evolved since the earliest days of broadcast television, and they have historically imposed many system parameters and constraints that ripple upstream to content creation systems. In order to explain these technical limitations and constraints and their origins, we will briefly discuss the history and evolution of Broadcast Networks and Television Stations, Cable Television, Satellite Television, Internet Video, Pre-Recorded Media (DVD/Blu-ray discs), and Digital Cinema and then describe the architecture typical to each of these systems today. Our task is complicated by the fact that technical developments among the various distribution systems are inherently intertwined – they did not evolve in isolation, but rather, technology innovations and developments crisscrossed industry sectors very rapidly, especially in the digital era as digital video and audio compression revolutionized content delivery and consumer electronics.

This brief overview of the history and evolution of distribution systems only serves the purpose of creating some basic understanding and appreciation of how certain technical parameters and limitations came about and how they ripple upstream to media creation systems and workflows. This chapter cannot possibly do justice to the complex history of many technical contributions and advances that took place over decades of innovation. Readers interested in the history of television technology development are encouraged to delve into more authoritative and historically accurate resources.

Similarly, this chapter cannot possibly do justice to the deep technical details associated with the various distribution systems. Again, we summarize the technical fundamentals of the systems in order to create a basic understanding and appreciation of the complexities and transformations that are inherent in getting media content distributed to consumers in so many technical diverse ways.

HISTORY AND EVOLUTION OF BROADCAST NETWORKS AND STATIONS

Experimental television developments were a hot technology topic of the 1930s and development efforts were taking place around the world. By the late 1930s, receiver manufacturers started to agree on early standards. In analog television systems and their cathode ray tube displays, it was extremely beneficial and economical to relate the key system parameters to the power line frequency. Hence, European efforts developed 50 Hz based systems while US efforts developed 60 Hz based systems – their respective power line frequencies. That early separation continues to this day and is the root cause for the variety of television standards that exist in the world today. Moreover, technology developers and government regulators of the time debated the amount of Radio Frequency spectrum that should be used for a television broadcast.

In the US, RCA's DAvid® Sarnoff famously proclaimed the advent of television at the 1939 World's Fair. The US Federal Communications Commission (FCC) had not yet allocated frequencies for television broadcasting and wanted a higher-resolution system. The National Television Systems Committee (NTSC) was convened, which in 1942 agreed upon the 525-line, 60 Hz monochrome NTSC television standard that used 6 MHz transmission channels. European developments resulted in 625-line, 50 Hz standards that used 7 or 8 MHz transmission channels. The commercial launch of television began in the post-war era.

The development of commercial broadcast television networks and stations followed the established model for radio broadcasting. Competing networks produced content (live shows) and distributed it to local television stations that were "affiliated" with a network. Originally, the distribution of network content to affiliated stations was accomplished by a terrestrial "long lines" telecom network. Like radio, there were some open times when locally originated shows were aired.

Even as monochrome television was being launched and adopted, the technology race to develop color television was underway. Color television requires that red, green, and blue images to be captured at a camera and then transmitted and emitted as red, green, and blue images by a display. Color television presented many new technical challenges, including how to display the red, green, and blue images that are needed to create a color picture, as well as how to transmit the RGB signals without simply requiring three times the transmission bandwidth of monochrome television. In the US, CBS developed an electromechanical approach that sequentially displayed red, green, and blue images using a rotating "color wheel" to sequentially create the colored images. The approach exhibited color fringing artifacts on moving objects, but it was approved for use by the FCC.

Other companies, including RCA, continued to pursue all-electronic approaches and the goal of developing a backward-compatible color TV system, with a transmitted signal that could be received and displayed by existing monochrome television receivers. These goals were met by the invention of the shadow-mask CRT (Cathode Ray Tube) and a signal format that mixed portions of the R, G, and B signals together in different portions (using a matrix multiplication) that are referred to as *component* signals.

A shadow mask CRT is a fine mesh structure that allows three parallel electron beams to independently energize emissive red, green, and blue phosphors. Color triads of phosphor dots formed what in modern terms we might describe as "color pixels." While subdividing the pixel structure of the image is fundamentally coarser than a monochrome display, the basic concept of "color sub-pixels" continues to be a fundamental construct in modern display technologies.

The transmitted signal problem was solved by understanding that an arithmetic combination of R, G, and B signals could form a "luminance" signal that approximates the signal created by a monochrome camera sensor. The luminance signal (technically referred to as the Y signal) was suitable for monochrome receivers, but a technique had to be found to send additional color information. A characteristic of linear signal combinations (a matrix multiplication) is that three signals can be created that have an inverse matrix transformation. This allowed R, G, and B signals to be transformed into "Y," "R-Y," and "B-Y" component signals.

Simply adding "Y" and "R-Y" recreates the red signal, while adding "Y" and "B-Y" recreates the blue signal. The green signal G is re-created by the inverse weighted average (matrix multiplication) of "Y," "R-Y," and "B-Y" component signals. Thus the "R-Y" and "B-Y" signals are technically referred to as "color difference" signals and denoted as Cr and Cb. The Y, Cr, and Cb component signals are thus collectively an alternate representation of an R, G, and B color image representation.

Now the challenge is that the Y, Cr, and Cb signals would apparently require three times the transmission bandwidth of a monochrome signal. Fortuitously, the human visual system's acuity (i.e., our ability to perceive sharpness) varies with color. It was found that the Cr and Cb component signals could be reduced in bandwidth compared to the luminance signal Y without causing a large loss of perceived sharpness. This fundamental use of different bandwidth component signals to represent color images is foundational – it continues to be the basis of modern component digital video signals, which are typically sampled in the Y:Cr:Cb ratio of 4:2:2 (i.e., meaning that the Cr and Cb signals are one half horizontal resolution compared to the Y luminance signal).

Finally, it was understood that the narrower bandwidth Cr and Cb signals could be modulated on a color subcarrier that was added to (i.e., superimposed on) the luminance Y signal. The resulting signal is referred to as a "composite" signal because it combines the Y, Cr, and Cb signals into one signal for transmission.

The color subcarrier was meaningless to (but only marginally visible) on existing monochrome receivers. Further, by making the color subcarrier frequency an odd multiple of half the line frequency, the pattern would not be static, but rather would "crawl" and be less visible on monochrome receivers. But a color television receiver could reconstruct the R, G, and B signals from the transmitted Y, Cr, and Cb component signals.

In the US, a second NTSC committee was convened, and it completed the NTSC color television standard in 1953. The NTSC standard maintained the basic signal parameters of 525 scan lines and a 6 MHz transmission channel, while adding new color subcarrier at a frequency of 3.58 MHz. Unfortunately, it was found that the color subcarrier caused interference with the audio subcarrier on some monochrome receivers. Rather than shifting the audio frequencies, this problem was solved by decreasing the video frame rate by 1 part in 1,000. Thus the 60 Hz monochrome television standard became 59.94 Hz (60 * 1,000/1,001) color television. While this decision has subsequently plagued modern television control and editing systems, it must be remembered that this decision was made long before video tape recording and editing were conceived. The development of European color television transmission standards PAL (Phase Alternating Lines) and SECAM took a conceptually similar path, but wider component signal bandwidths could be used in their 7 and 8 MHz transmission channels, while other improvements were made and the precise 50 Hz field rate was able to be maintained.

The new all-electronic, backward compatible NTSC color system was approved by the FCC and it replaced the early color-wheel approach. Color television broadcasts began thereafter, but color television cameras, production equipment, and receivers were relatively expensive and complex. Both network and local station facilities used the composite signal format (i.e., the same format as the transmitted signal) as the single-wire interface for equipment interconnection and signal switching.

Color broadcasts grew slowly, and it was not until 1964 that the RCA-owned NBC network became the first network to have all of its content in color. Meanwhile, technical advances continued to reduce the complexity and cost of composite signal interfaces, switching and recording, which typically remained analog well into the 1980s. Adding another media type to basic picture and sound elements, the development of closed captioning for hearing-impaired consumers in the mid-1970s made use of line 21 of the NTSC analog system to carry digital data for the characters of the caption text, which was specified in the CTA-608 technical standard.

In the early 1980s, the technology to digitize video signals started to become practical. Initial digital standards for video simply involved digitizing the analog composite signals using a sampling rate that was an integer multiple of the color subcarrier frequency. With the realization that the individual component signals were needed for many of the processing steps in the content creation and broadcast infrastructure, there was much international discussion about component digital video and the sampling rates for both 59.94 and 50 Hz systems. The industry rallied to test these concepts and the SMPTE Component Digital Video demonstrations in 1981 established the viability of using sampling rates that were unrelated to color subcarrier frequencies of analog transmissions. A 13.5 MHz sampling rate was a "magic" frequency that was an integer multiple of the horizontal line frequencies for both 59.94 and 50 Hz systems. ITU-R BT.601 established those parameters and it remains the foundation of the resulting Serial Digital Interface (SDI) interface standards as well as the foundation for resolution (pixel formats) in both HDTV production and signal interface standards such as HD-SDI, as well as modern digital television transmission standards.

By the mid-1980s, work on HDTV was well underway all around the world. Doubling the horizontal and vertical sharpness of traditional television for HDTV and increasing the picture aspect ratio from 4:3 to widescreen 16:9 required five times as much signal bandwidth. This would be difficult enough within a production and studio facility, but the problems of transmission systems were once again the tremendously difficult challenge. Researchers sought to extend established analog color television techniques and to develop new temporal techniques that combined information across several frames of video. Clever system approaches such as MUSE (Multiple SubNyquist Encoding) from Japan and HD-MAC (High Definition Multiplexed Analog Components) from Europe demonstrated that HDTV transmission might be feasible using two traditional television RF channels.

Spurred by research efforts on HDTV around the world, in 1987 Broadcasters petitioned the FCC to establish a new advanced television system for the US. In the ensuing race, 23 analog system proposals were put forth, but none could deliver full HDTV in a single 6 MHz RF channel. By 1990, four radically new all-digital HDTV system proposals emerged. After rigorous testing, a Grand Alliance was formed to forge a "best of the best" combination, which became ATSC[1] (1.0) – the world's first digital television standard.

The ATSC system provided broadcasters with the ability to deliver high-quality, widescreen HDTV in either 1080i or 720p format, with 5.1 surround-sound audio, using digital transmissions that were free of analog noise and ghosting degradations. It also offered broadcasters the ability to send multiple program streams and "multicast channels" of standard definition (480i and 480p) video and stereo audio along with their primary HD programming. The multi-stream capability also enabled the emergence of Video Description audio services for visually impaired consumers. Other digital TV standards soon followed in other regions of the world, notably the DVB (Digital Video Broadcasting) and ISDB-T (Integrated Services Digital Broadcasting – Terrestrial) systems.

HDTV receivers were introduced in the US in 1998 and their adoption by consumers coincided with the availability of large flat-panel displays. By 2009, the last full-power analog television stations were shut down in the US. In 2012, the ATSC began work on the next-gen ATSC 3.0 standard that will deliver UltraHD (4K) and High Dynamic Range (HDR) signals, high frame rate formats, immersive audio, and personalized and interactive consumer experiences. ATSC 3.0 was approved by the FCC in 2017, and as of this writing in 2020, the commercial launch of ATSC 3.0 is beginning in the US and is well underway in Korea.

MODERN BROADCAST NETWORKS AND STATIONS

Referring to Figure 6.1, today's broadcast networks aggregate content from both scripted (pre-recorded) and live productions (notably sports and news). The details of both types of productions are described in detail in the Production chapter of this book.

Live productions typically provide a contribution feed to the Network Operations Center (NOC). Typically, this is a fiber connection with a very high quality, high bit rate compression such as a J2K codec at ~200 Mbps. C-band satellite feeds are also often used for backup, using a more highly compressed, lower bitrate format such as H.264 at ~30 Mbps. At the Network Operations Center, the contribution feeds are decompressed into baseband video signals (e.g., HD-SDI).

Scripted productions and advertisements typically provide a high-quality file transfer to the NOC.

Network signals containing picture, sound, closed captioning, and video descriptive services are usually distributed to Affiliate stations over digital satellite links. Broadcast networks in the US generally have different feeds for different time zones, and some networks provide additional programming versions that are tailored to other geographic and/or audience interests (e.g., a sports match that is of interest in the cities of the two competing teams). The assignment of signals to various satellite transponders and the control of the corresponding reception at Affiliate stations is typically controlled by a Network Operations Center (NOC). The network signals are typically compressed at higher bit rates than the eventual transmission by an Affiliate station. H.264 and MPEG-2 are the most common video codecs for network signal distribution.

Referring to Figure 6.2, at a typical Affiliate Station, the satellite receive dishes and decoders are pointed at the proper satellite and tuned to the proper transponder to receive a network feed for the proper time zone (and sometimes geographic location). The most sophisticated network operations systems automate this process and may deliver many network "versions" to appropriate groups of receiving stations. Simpler systems deliver time-zone-based feeds and may rely on the local station to properly point their receive dish and tune to the proper transponder. An IRD (integrated receiver decoder) at the station decompresses the network feed, resulting in the network signal feed on an HD-SDI interface.

Television station infrastructure is typically built around an HD-SDI switching core, which also includes local production capabilities for both live and scripted content. Stations' playout systems are used for syndicated and locally generated content. The station's Master Control provides for switching between network and local content and is the final quality control point before the actual broadcast transmission of the signal.

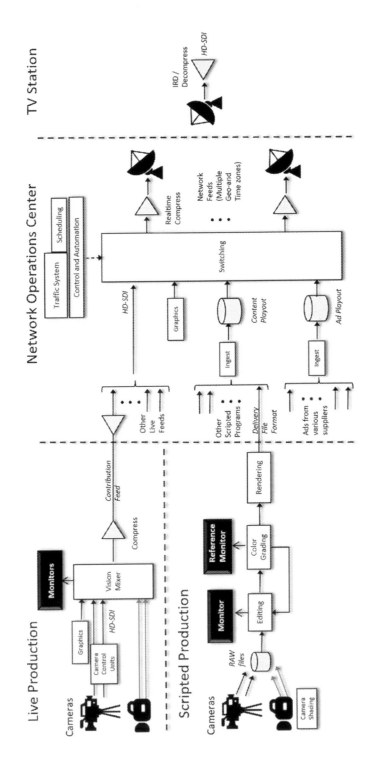

FIGURE 6.1 Broadcast Network Content Workflow

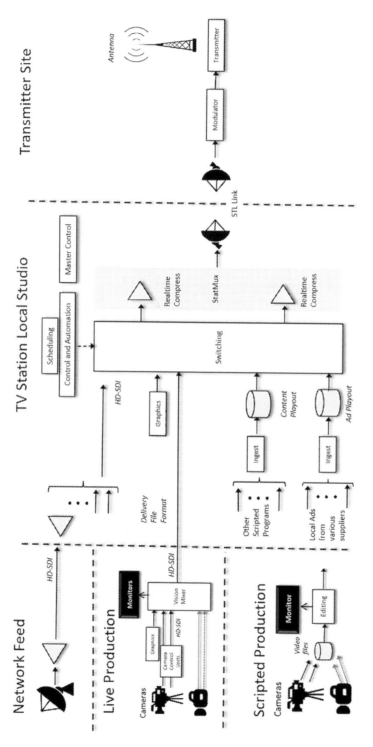

FIGURE 6.2 TV Station Content Workflow

HISTORY AND EVOLUTION OF CABLE SYSTEMS AND CABLE CHANNELS

Cable television began to emerge as early as the 1960s, as a means of receiving broadcast television signals in areas with poor signal reception conditions. In such situations, a Community Antenna Television (CATV) often consisted of a tall TV antenna, with amplifiers driving the received broadcast signal over coaxial cable to homes in the community. Early cable systems provided better signal reception and picture quality than might otherwise be available to a consumer with a home antenna. Cable systems proliferated and their capacity to carry many channels of video also grew, as cable and amplifier quality (low-noise) and bandwidth capability advanced.

Cable systems eventually began to have more than enough capacity to carry all of the local broadcast television signals. The excess capacity was filled with the advent of specialized "cable channels" (e.g., HBO, MTV, etc.). Early cable channels used C-band analog satellite distribution to the many individual cable system operators' headends. The cable systems themselves used the same transmitted signal format as broadcast signals, modulated to a different Radio Frequency (RF) channel on the cable. "Cable-ready" televisions provided the extended tuning capability and the coaxial "F-connector" that made consumer adoption of cable television simple and convenient.

As the cable industry expanded, consolidations began to occur. Companies that operated systems in different communities became referred to as Multiple System Operators (MSOs). The demand for channel capacity continued to increase as many cable channels continued to find audiences and provide unique content niches. Technology advances resulted in the development of Hybrid Fiber Coax (HFC) systems. The use of fiber links to local neighborhood nodes reduced the number of cascaded amplifiers and allowed further increases in capacity. By the early 1990s, many cable systems had between 350 and 500 MHz bandwidth, providing some 50–80 channels of analog television. Cable systems became increasingly sophisticated and the insertion of local advertising into cable channels became an established business. The initial technology relied on "cue tones" carried in the audio to trigger an insertion event. Eventually, this also became possible for the replacement of ads in local broadcast signals as well.

In the late 1990s, the digital television technology that revolutionized broadcast was poised to similarly impact cable television, particularly since industry representative had participated directly in the testing and evaluation of the Grand Alliance digital HDTV that was the basis for the ATSC broadcast standard. The same MPEG-2 digital video compression and transport was directly applicable to cable. However, the Vestigial Sideband (VSB) modulation and transmission approach that has some broadcast-specific innovations was not necessary for cable. Its very closely related and more traditional Quadrature Amplitude Modulation (QAM) was found to be suitable for a 6 MHz RF channel on a cable system. Digital standard definition cable channels significantly improved cable picture and sound quality and became widely deployed by the cable industry in the US in the early 2000s. At that time, a typical cable system had its "analog tier" in the lower frequency range of 6 MHz RF channels (e.g. 350 MHz and below) and its "digital tier" of QAM signals at higher frequencies. This approach allowed cable operators to continue their legacy analog services with no disruption to viewers, while launching the new tier of digital services as a premium service.

The advent of digital channels on cable was also inherently a catalyst for the advent of on-demand content. The earliest Video On Demand (VOD) approaches located video playout servers in proximity to a QAM modulator and assigned the requested content stream to a vacant QAM channel. This approach was costly to scale, since it required many replicated video server systems.

At the same time, the large data capacity of QAM digital video channels was apparent and Internet use was growing rapidly. CableLabs developed techniques for upstream data transmission on cable systems and produced the first Data Over Cable Service Interface Specification (DOCSIS) "cable modem" spec in 1997. Early DOCSIS cable modems could deliver 40 Mbps data speeds with a 6 MHz QAM signal. The deployment of DOCSIS cable modems drove an overhaul of cable for upstream communications, wider bandwidth downstream capability and the deployment of Hybrid Fiber Coax (HFC) architectures that use high capacity fiber delivery to local neighborhood nodes (typically 200–500 homes) and then convert to traditional coax cable for the "last mile" delivery to the home.

By the mid-2000s, subsequent versions of DOCSIS brought higher upstream data speeds, enabled voice-over-IP (VOIP) capabilities, and introduced IP v6 support. The cable industry deployment enabled the "triple-play" of video, voice, and data services. At that point a cable system consisted of three "tiers" of 6 MHz channels – the legacy analog television tier, a digital television QAM tier that delivered both HD and SD video programming, and a DOCSIS tier that carried digital telephony and Internet data. Cable operators served their set-top boxes with video services and provided gateway devices (i.e., DOCSIS modems) that provided Internet and telephone services. As WiFi became ubiquitous in consumer PCs, tablets, and smart phones, cable gateways provided the connection to the Internet and the WiFi hub for the home. In 2006, DOCSIS 3.0 provided "channel bonding" capability that could logically combine multiple downstream QAM channels to multiply downstream data speeds, achieving up to 1 Gbps downstream data speeds.

In 2013, the DOCSIS 3.1 spec allowed for different downstream data channel bandwidths, finally separating cable data transmission from the original 6 MHZ analog video channelization of the 1942 NTSC standard. With DOCSIS 3.1, 10 Gbps downstream speeds could be theoretically achieved. Throughout the 2010s, cable data service speeds continued to grow and become more widely available to consumers and businesses. Most recently, the 2017 DOCSIS 4.0 spec now allows fully symmetric data speeds and it is just beginning commercial deployments.

TRADITIONAL CABLE SYSTEMS AND CABLE CHANNELS

Before we progress to a description of modern cable systems and their Internet and video streaming capabilities, it will be illustrative to examine a "traditional" cable plant that simply delivers linear television channels. This would be typical of a cable system in the mid-2000s.

Figure 6.3 provides a very simplified high-level view of the end-to-end content ecosystem for the case of live linear channels. The Figure is intended to provide a high-level reference for discussing the functionality of the programmer and cable operator ecosystem; it is not representative of specific systems or architectures, which can differ among various content producers, programmers, and cable operators.

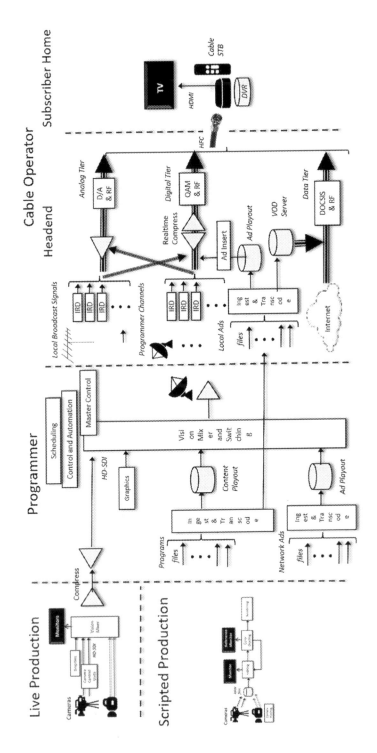

FIGURE 6.3 Traditional Cable/Linear Channels

Referring to Figure 6.3, on the 'production' side of content, it is important to understand that there are two categories of linear content creation and distribution – truly "live television" streams, which may include sports, special events, news and other content, and linear streams which are a compilation of pre-recorded content (a traditional linear television channel).

The linear television signals that cable operators deliver to their subscribers come from many sources, including cable Programmers, Broadcast Network's local TV Station Affiliates and independent local broadcast TV stations, and operator-produced programming such as sports and news. These feeds may be in HD (16:9 1080i or 720p) and/or SD (4:3 480i) formats, which are delivered to consumers on different "tiers" of the cable system, which may include analog, digital standard definition, and digital high definition. These tiers require different consumer equipment and cable Set-Top Boxes. As of this writing, the analog video tier on cable is quickly disappearing, although it persists in many small systems in the US and internationally in emerging economies. And although virtually all modern TV displays are HD, many secondary cable STBs are still only SD-capable, pending their slow but inevitable replacement by HD boxes. Thus, for the foreseeable future, programmers and content owners must deliver content to cable operators in both HD and SD formats, or allow them to down-convert HD to SD using Active Format Description (AFD) metadata.

Referring again to Figure 6.3, as a live production signal is fed to a Cable Programmer, advertisements are inserted at appropriate points during the event. This is accomplished by ad playout servers and the same live signal interfaces that are used in production. Therefore, current operational practice is constrained to require that the ad inventory use the identical video format to the live production. Programmers may also insert graphics elements "downstream" from the actual production, often including channel branding logos and promotional animated graphic elements. Programmers then distribute their fully integrated linear signal to cable operators and other distributors, often applying video compression during this step.

Cable Operators receive linear channel feeds from many different Programmers. In addition, cable operators re-transmit local TV broadcast stations. In many larger markets, a direct fiber connection is available between TV stations and cable ingest points. However, many markets continue to rely on direct reception of broadcasters' over-the-air signals as the source for re-transmission.

Cable operators also have a significant business in ad insertion for both Programmer and Broadcast channels, since they can perform geographic and/or demographically targeted ad insertion. Advertising segments used by cable operators come from a wide variety of national and local sources. Cable operators' ad insertion systems and ad inventories must deal with the variety of source signals and video formats (e.g., HD 1080i, HD 720p, SD that are distributed by various Programmers and Broadcasters). Moreover, cable operators must be able to insert advertising in linear channels, "start-over" playback, DVR playback, and on-demand viewing. The technical quality, complexity, and latency must meet both operational and business agreement needs. Note that the adoption of multiple formats affects the implementation scale in terms of transmission and storage (including ads).

In the case of delivery to a cable-operator controlled set-top box (STB), video compression at the headend and decompression in the STB are controlled by the cable operator, as are the insertion of navigational and closed caption graphics. The consistency and visual quality of navigation functions (e.g., picture-in-picture, multi-channel mosaic screens, etc.)

and graphics across various programming channels are important requirements for providing a high-quality user interface and navigation experience to subscribers.

Cable STBs connect to modern TVs with the digital HDMI interface. Although analog interfaces still exist in legacy STBs and old TVs, typically only HDMI with content protection such as HDCP (High Definition Content Protection) is used for HD content. The video format of an HDMI connection is determined by a special "handshake," during which no video data is transferred. The viewer disruption associated with an HDMI format switch has resulted in current practice being to have the STB output a single video format to the TV. This means that regardless of the actual transmission format, resolution conversions are often performed by the STB hardware.

HISTORY AND EVOLUTION OF SATELLITE TELEVISION

Even as cable television adoption grew, the capability to distribute television over a large geographic footprint was an opportunity that was potentially well-suited for satellite television distribution. For a time in the 1980s and 90s, an illegal market for large backyard C-Band receive dishes existed. Consumers with a large dish could (illegally) receive the cable channels that were being sent to cable headend without paying to receive the services. This started a cycle of applying analog scrambling techniques to make the signals non-standard, but black-market descramblers became readily available. By the late 1980s, this situation started to motivate the conversion of cable channel programming to cable headends to move to a digital format, where digital encryption could be applied as a strong content protection measure.

In the early 1980s, much work was underway around the world to explore the opportunities of direct-to-home (DTH) satellite television. High power satellites were being developed, which would help to reduce the size of consumer receive dishes to practical sizes for use in typical homes. The World Radio Conference designed frequencies and orbital slots for Direct Satellite Broadcasting (DBS).

Attempts were also made to improve the quality of television delivered over satellite, so analog component signals (Multiplexed Analog Components) were defined that eliminated some of the artifacts of composite video signals (e.g., NTSC and PAL). Although they had some limited commercial success in some areas of the world, analog satellite systems had far less capacity (typically 24 channels) than satellite and were more costly for both operators and consumers.

After the four digital systems had emerged in the US race for digital HDTV, it became apparent that the same technology could be applied to satellite television. With the efficiencies of digital compression and the robustness of digital transmission, satellites could deliver amounts of programming that was on par with cable systems of the time. DirecTV was formed and launched the world's first Digital Satellite System in 1995, using MPEG-2 compression and the proprietary data packet format of the Advanced Digital HDTV system. Satellite set-top boxes quickly incorporated disc storage for Digital Video Recorder (DVR) consumer recording and to provide Video On Demand Services. Popular VOD movies are highly compressed and sent as data files to the DVR disk, which serves as a data cache for on-demand content. Able to provide pay-TV services to rural areas of the US that were not reached by cable, adoption grew rapidly (at the time, it was the fastest-adopted consumer product in history; only later surpassed by DVD). DISH Network followed suit and as of this writing, the two satellite services reach over 25M subscriber households in the US.

DTH (Direct To Home) satellite television systems have continued to advance their digital technology base since their initial launch with MPEG-2 compression and standard definition video. In the early 2000s, DirecTV began offering HDTV services, which shocked the US cable industry, where many thought that digital standard definition would satisfy consumers. DTH satellite systems migrated to the next-generation MPEG-4 codec and high-definition video and are currently in the process of introducing the H.265 codec and Ultra High Definition (4K) video.

The DVB standards organization developed successive generations of standards for satellite television (DVB-S, DVB-S2, and DVB-S3), which helped to catalyze the launch of DBS systems around the world. As of this writing Satellite TV is a widely adopted distribution technology in most areas of the world.

As broadband Internet connectivity became increasingly common, satellite STBs incorporated Ethernet and WiFi connectivity, enabling operators to provide content via over-the-top streaming in addition to linear video channels on the satellite and locally cached content.

MODERN SATELLITE TELEVISION SYSTEMS

Today, a typical DTH satellite system is often a hybrid of satellite and over-the-top infrastructure. The satellite system operator gathers its many channels of linear programming from a combination of C-band and KU-band satellite receive dishes and IRDs (the same satellites, transponders, and signal that also serve cable operators) and direct fiber connections to large media companies' distribution centers.

In areas where local TV station signals are retransmitted on pay-TV satellite systems, the local station's signal is either directly fiber connected to the DBS operator, or alternatively, all of the stations are received from an over-the-air antenna and connected to the DBS operator on a fiber. At the satellite system operator's Network Control Center, the video feeds are converted to baseband signals, and efficiently compressed into satellite transponder sized multiplexes, using Statistical multiplexing.

HISTORY AND EVOLUTION OF INTERNET VIDEO AND OTT SERVICES

Internet Protocols (TCP/IP) were invented in the 1960s and became widely used to interconnect large computers in academia and industry. The advent of personal computers created an environment for dial-up modems to connect them to "Bulletin-board" services where they could interact and share information among users in simple ASCII text format. Early bulletin board services like CompuServe and AOL were launched in the mid-1980s. The IBM PC and Microsoft DOS sparked the introduction of PCs into the corporate world, driven by applications such as word processing and spreadsheets. Local area networks (Ethernet) enabled corporate PCs to share files and utilize email and Internet connections bridged corporate campuses, and interconnected with academic computing centers. Graphic User Interfaces (GUIs), windows, and the mouse were inventions of the late 1990s that greatly increased the usability of personal computers.

By the early 1990s, Digital HDTV system developments were already underway when HTTP and the browser were invented. The Grand Alliance HDTV system was nearing completion when Windows 95 was released and brought the windows and mouse innovations to PC users. Even as digital HDTV was on the verge of launch, there was debate about the future roles of televisions and computers; phones were analog and 1G analog cell phones were the size of a shoebox. Consumer Internet access was predominantly dial-up modems, which had increased in speed from 1.2 kbps to subsequent 2.4, 4.8, 9.6, and 19.2 kbps generations. Modems over telephone lines were reaching a fundamental limitation in performance, but higher speeds were becoming a necessity as relatively data-hungry HTML pages began to be consumed by browsers. The innovation of Digital Subscriber Lines (DSL) from phone companies enabled higher speeds and direct connection to an Internet "home page," but they were expensive and only available in limited geographic areas. Nevertheless, pioneering developers began to experiment with video delivery on the Internet. While quality was significantly poorer than standard definition video that was available on cable and satellite, the integration of video content on web pages began.

As described in the previous section, the growing deployment and adoption of Cable Modems (DOCSIS) in the early 2000s brought high-speed Internet connectivity to a rapidly growing number of homes. With higher data speeds, the quality of video increased commensurately and the opportunities for video content delivery on the Internet became more apparent. Notably, YouTube was founded in 2005 and was quickly acquired by Google in 2006.

Unlike web pages and data files, the delivery of video requires a consistent "stream" of data to be delivered. Internet Protocols were not designed for such an application, being inherently a "best effort" approach to communications over a network. The quality of streaming video delivery was significantly advanced by three technical innovations. First, Content Delivery Networks (CDNs) were invented to cache Internet content assets at the "edge" of the Internet cloud, geographically closer and fewer network communication hops to users. CDNs accelerated the delivery of all web page content assets, but they were crucial to achieving video streams. Second, the concept of Adaptive Streaming is that there can be multiple representations of a piece of content stored as different files having different resolutions and data rates. If the nominal data rate of a video stream cannot be maintained, the video player can temporarily access a lower quality, lower bit rate version of the content. This is important at the start of content playback, in order to reduce long wait times while the decode buffer is initially being filled (i.e., the familiar "buffering" message"), as well as to avoid the stream totally freezing due to intermittent network congestion. Third, the development of HTML5 (the latest version of web page standards) provides for a <video> tag, in which video elements are easily represented on web pages. Furthermore, its Encrypted Media Extensions (EME) allow for content encryption and protection by Digital Rights Management (DRM) systems, enabling pay-TV subscription models. While proprietary DRM systems existed since the 1990s, their seamless integration into the web was significantly advanced by HTML5.

Although it was initially launched as several proprietary, non-compatible approaches, adaptive streaming technology has matured and become more standardized. Dynamic

Adaptive Streaming over HTTP (DASH) is a standard developed in the MPEG group and it is gaining widespread adoption. However, Apple continues to support their HTTP Live Streaming (HLS). As of this writing, DASH and HLS are the two most common video streaming protocols.

Similarly, encryption and DRM systems for content protection began as highly proprietary systems. The approach of Common Encryption (CENC) has been standardized and it allows the use of multiple DRM systems to protect content. As of this writing, Google Widevine, Microsoft PlayReady, and Apple Fairplay are among the DRM systems that are in widespread use.

In fact, seeking compatibility with Internet video standards, the new ATSC 3.0 broadcast standard makes use of HTML5, DRM systems, and common encryption Internet standards.

Thus, using the streaming video standards of the modern Internet and the high data speeds of DOCSIS cable modems, video content can be delivered "over-the-top" of traditional cable systems and their legacy video-centric technical standards.

MODERN INTERNET VIDEO AND OTT SERVICES

Today there are many Over-the-Top (OTT) video services available as either subscriptions or ad-supported content. These services can be accessed via a web browser, but the proliferation of smart phones, tablets, smart TVs, and streaming video set-top-boxes has given rise to service-specific "Apps" (applications) for those devices. Most OTT services provide on-demand content, but live channels are increasingly becoming available on some services (Figure 6.4).

An OTT service operator ingests content and applies video compression encoding to produce multiple resolution and quality levels for its adaptive streaming "stack" of versions that are most appropriate for different connection bitrates. Additional versions may be performed to support different video codecs that are supported by various consumer devices. At the same time, the video is broken into short segments of a fixed time duration, typically in the range of 2–10 seconds. Each of these segments is a file. The video data for a given piece of content is thus a large collection of files that are all of the segments for all of the adaptive bit rate stack, which is represented by a master Manifest file.

The complete set of files for each piece of content is published on an "origin server." From there, copies are propagated throughout one or more CDN systems, so that the data resides in close physical and connection proximity to users.

When a user App (or web browser) requests a piece of content, the OTT service's web server determines the user device's capabilities, supported protocols, and DRMs and the best video resolution and codec that is supported by the device. A session-specific Manifest is created that contains the nearest CDN cache's URLs (i.e., file names) of each ABR representation for each time segment is built and sent to the App. Content playback occurs by the App requesting each successive segment of content from the CDN. As the data is retrieved from storage a "Packager" wraps the compressed video data in the appropriate protocol (e.g., DASH or HLS) and provides the decryption keys for the appropriate DRM system.

At the user device, the requested data fills a buffer and is subsequently decompressed into a baseband video representation for display on the device. The App monitors the buffer state and if it is too low, it indicates that the Internet data rate cannot keep up with the video

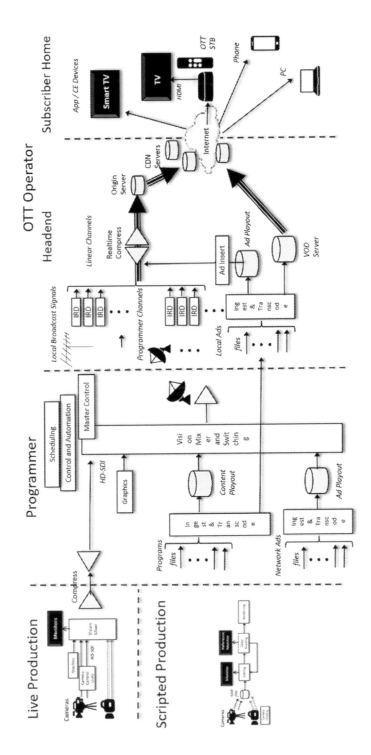

FIGURE 6.4 OTT Content Workflow

decoding rate. In such a situation, the App will examine the Manifest file and request its next segment from a lower level of the ABR stack. By continually monitoring the device's buffer state, the App can "downshift" to a lower quality / lower bitrate representation during temporary period of network congestion and "upshift" back to a higher quality / higher bitrate representation as the Internet connection speed improves.

Digital ad insertion (DAI) for targeted advertising is one of the capabilities of OTT that is often utilized. The simplest OTT systems create the Manifest with URLs for the desired ads, which are retrieved by the App (client software) in the course of content playback. More sophisticated handshakes are used to trigger an event at a commercial break, at which time the App queries an Ad Decision Server to get the URLs of the target ad, content, which are retried by the App. This approach is referred to as "client-side ad insertion." Alternatively, OTT systems increasingly employ "server side ad insertion," where the CDN issues a new manifest to the App.

MODERN CABLE SYSTEMS

Modern cable systems are an amalgamation of traditional cable and OTT architectures. In addition to live linear transmission and on-demand content, cable operators usually provide DVR, "cloud DVR," and "start-over viewing" functionality to subscribers. Modern cable systems deliver traditional TV services to operator-controlled set-top-boxes (STBs) and "TV Everywhere" services to Internet-connected consumer devices with an operator App. Many cable system operators acknowledge that they are beginning a transition to an all IP network and operations infrastructure. In support of this, many cable STBs contain traditional MPEG-2 video and QAM modulation decoding as well as a DOCSIS modem and support for additional video codecs and protocols used in OTT streaming. This will enable the migration of on-demand, cloud DVR, and startover services from cable-industry-specific systems to solutions that are based on commodity IP and Internet hardware and software for streaming video. However, it is important to understand that the details of network utilization, STB software, and the level of support for streaming video delivery to various consumer devices to vary widely among cable operators.

Today, the digital content and ad files delivered to cable operators are transcoded to the specific formats required by their client STBs to receive <u>cable</u> on-demand and OTT on-demand video streams.

In addition, many on-demand services require the insertion of advertising segments, which similarly come from a wide variety of sources. These on-demand systems and the insertion of advertising require that the same video format is maintained across the content and the ad.

The considerations involving consumer devices and interfaces are essentially the same regardless of whether linear or on-demand content is being delivered to a subscriber.

HISTORY AND EVOLUTION OF PRE-RECORDED MEDIA

With the emergence of consumer video tape cassette recorders in the late 1970s, prerecorded cassettes of movies became immensely popular for consumer rental and purchase. A protracted "format war" was waged between BetaMax and VHS formats, with VHS being the ultimate commercial winner. The analog tape standards were lower quality than analog broadcast and cable delivery, but the convenient availability of content was new and compelling for consumers.

During the 1980s, technologies for higher-quality pre-recorded analog discs were developed using both capacitive and laser playback technologies. The formats did not achieve widespread consumer adoption, but they paved the way for the next generation of digital technology that would come in the 1990s.

Following the development of digital television in the early 1990s and virtually in parallel with the commercial development of digital satellite television, the pre-recorded digital optical disc DVD (Digital Versatile Disc) standard was also enabled by MPEG-2 video compression and advances in optical recording density, which could store approximately 8.5 GBytes on the same 12 cm disc form factor as audio CDs. The use of multiple layers can be used to further increase storage capacity.

DVD was launched in the mid-1990s. Before commercialization, competing factions came together to avoid another 'format war" and DVD was a huge success that enjoyed rapid consumer adoption (at the time it was the most quickly adopted consumer product in history). Despite its age and technical obsolescence, DVD distribution still remains in the media landscape today.

The DVD format provided conventional 4:3 aspect ratio or widescreen 16:9 standard definition digital video, along with 5.1 channel surround sound, closed captions, and subtitle capabilities. With MPEG-2 video compression at typical bit rates of about 6 Mbps for 480p24 (film) and 480i (video) formats and Dolby Digital and other surround soundtracks, DVDs provided consumers with a huge leap in the picture and sound quality and convenience of pre-recorded content. It had a simple menu system that enabled chapters for simple consumer navigation of content. Many movie DVDs contain bonus features and extras to help entice consumer purchase of the content. Specialized DVD authoring tools are required to create discs, which were initially manufactured only in large optical disc pressing plants. The advent of recordable and re-writable DVD discs (+R, –R, and –RW formats) unleashed the potential for consumers to easily edit and create their own home videos using a camcorder and a personal computer.

By the mid-2000s, HDTV adoption was growing, and it became obvious that a successor optical format would be needed for HDTV content. By the early 2000s, an industry "format war" erupted between the competing HD-DVD and Blu-ray formats, which were launched commercially in the mid-2000s. Although Blu-ray eventually emerged as the winning format, the protracted industry battles and consumer confusion delayed the adoption of the format so that it quickly became challenged by emerging downloadable digital file formats and online/on-demand streaming delivery.

The Blu-ray Disc (BD) format uses the AVC/H.264 video compression standard (the successor to MPEG-2) and a blue laser that achieves higher data storage density – approximately 25 GBytes of data on the same size 12 cm disc as a DVD. Single, dual, three-, and four-layer discs are available. With these storage densities and the advances of AVC/H.264 video compression and advanced audio codecs, the Blu-ray format provides outstanding HDTV (1080p24 film and 1080i or 720p video formats) picture and surround sound quality to consumers. Like DVD, Blu-ray has menus and use navigation software framework, which in its technical details is based on the Java programming language that is much more modern than its primitive DVD predecessor.

The BD format has been upgraded and evolved to include 3D video and "Mastered in 4K" video that is delivered in 1080p format on the disc. In 2015, the UltraHD Blu-ray format was launched, which is capable of UHD (4K) video playback in both standard and High Dynamic Range formats and advanced immersive audio (object-based sound) capabilities. UltraHD Blu-ray players also support conventional Blu-ray and DVD playback.

Although UltraHD Blu-ray arguably provides the best-available video and audio quality for consumers, many industry observers speculate that it may be the last physical media format, being eclipsed by the convenience of online on-demand video streaming. Nevertheless, in the immediately foreseeable future UltraHD Bluray, Blu-ray, and DVD discs remain part of the media delivery landscape.

HISTORY AND EVOLUTION OF DIGITAL CINEMA

At the time when the development of digital HDTV for US broadcasting was progressing in the mid-1990s, cinemas continued to rely on the distribution of 35mm film and the use of film projectors in theaters. Several early attempts to apply digital HDTV technology to the cinema ecosystem were commercially unsuccessful but created an industry awareness of digital compression and file delivery technologies.

Digital Cinema Initiatives (DCI), LLC, a joint venture of the major film-making studios, was formed in March of 2002 with an objective of establishing voluntary specifications for digital cinema. A key goal was that the image and audio would minimally be equivalent to or better than a 35mm film print. Significant testing was accomplished and the first Digital Cinema System Specification (DCSS) was first published on July 20, 2005.[2] This was followed by the DCSS Compliance Test Plan on October 16, 2007, which was used by manufacturers to demonstrate compliance with the specification. While equivalent to or better than a 35mm film print, this effort brought about two other major advantages for the movie-going public: each presentation provided clean images, since there were no longer film prints gathering dust and getting scratched with each passage of film through the mechanical system of a film projector, and each presentation was steady, since there was no film print weaving/juddering through the mechanical system of a film projector. The changeover of film projectors to digital projectors began in earnest around 2005–2006, and by 2013 had occupied well over 90% of the commercial market in the United States.[3] Today, film projectors occupy only about 0.5% of the US market.

The DCI DCSS was adopted by and developed into approximately 50 fully detailed standards, recommended practices, and engineering guidelines by the Society of Motion Picture and Television Engineers (SMPTE).[4] All the key SMPTE standards needed for program interchange were later adopted and published by the International Standards Organization (ISO). The following sections will discuss the current state of these standards for distribution of D-Cinema content to movie theaters and will discuss the workflows and distribution architectures for both movie content and other live content to movie theaters. Figure 6.5[5] is the Digital Cinema System Workflow as currently presented in the DCI DCSS. The left green boxes describe the mastering of 2K or 4K movies and all the other assets as discussed earlier in this book. The turquoise blocks refer to the steps of creating the package, known

Digital Cinema System Workflow

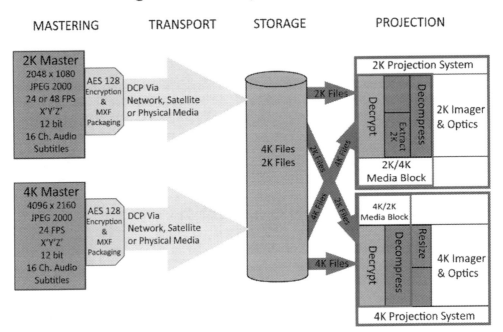

FIGURE 6.5 DCI Digital Cinema System Workflow

as the Digital Cinema Package (DCP) that will be sent to theaters. The blue blocks refer to the distribution of those DCPs to theaters. The yellow storage container refers to the theater storage system. The red arrows clarify that 4K movies can be played out by either a 2K or 4K projection system as shown by the boxes on the right. These items will be covered in more depth in the following sections.

MODERN DIGITAL CINEMA WORKFLOWS

The distribution of digital cinema content is unique from other forms of audio-video content in that the picture is not bound to the sound for distribution. Instead, a DCP is delivered to each movie projector that minimally contains two .mxf "track files" of picture and audio, and may optionally contain additional track files of subtitles, captions, etc., which the projector will select and render at show time based upon a unique Composition Play List (CPL) for that show. This way, a single DCP package can be distributed to many movie theaters, and each projector will select the files appropriate for that showing, and play all the appropriate pieces based upon the instructions in the CPL. This not only streamlines the distribution process by allowing all components to go to all theaters, but also provides more flexibility for the theater to customize their presentations to their audiences. As an example, a Paris theater showing a popular English PG film may play a version of the film that is dubbed in French for their matinee shows where young children who do not read may be present. But in their later evening hours, they may show the native English soundtrack and display French subtitles for an older audience.

Previous sections of this book have described the post-production aspects of creating the raw essence components that are needed for a movie presentation. Picking up from there, the following sections will describe the workflow shown in Figure 6.5. One item of note with respect to the raw assets is that while the bulk of the work is done in post-production to create the basic image and audio assets, more and more of the work done for "localization" has been pushed out from the post houses to the distribution houses. Previous chapters of this book have addressed the main localization items of dubbing in foreign languages, subtitling in the local language when a foreign language is being spoken in the movie, and closed captions in the local language for the hearing impaired. But there are also other miscellaneous localization tasks that need to be performed as well. Common to all films is converting the main opening movie titles and the credits into local languages. Another item that arises frequently is when an image with words appears in the film that needs to be localized to the country. As an example, imagine a children's animated film that has a STOP sign for the English-speaking markets. Using Computer-Generated Imagery (CGI), the word STOP will be changed to the appropriate signage word for each country.

Mastering

Mastering, as shown in the green boxes on the left of Figure 6.5, is the step of preparing the raw assets for distribution. In the following sections, the various track files, also known as "essence assets," which need to be prepared for the DCP, will be explained: Picture, Stereo (3-dimensional 3D) Picture, Sound, Immersive Audio, Captions and Subtitles, and Auxiliary Data. And there are always exceptions to the rules, captured in the Other Unique Formats section.

Picture

A picture track file must be included in a DCP. As mentioned earlier in the book, the aspect ratio (AR) of most movies are "flat" with an AR of 1.85:1 or "scope" with an AR of 2.39:1. Recognizing that theaters may also want to present modern television content, particularly for a pre-recorded live event, the modern HDTV AR of 16:9 or 1.78:1 was taken into consideration. DCI established a picture "container" of 2048 pixels wide by 1080 high for a 2K distribution and 4096×2160 for a 4K distribution, in which all these ARs could be handled. From these specifications, a scope 2K movie will use 2048 pixels wide by 858 high and a flat 2K movie will use 1998 pixels wide by 1080 pixels high ratio.

After significant testing, DCI selected JPEG2000[6] for picture compression at a maximum bit rate of 250 megabits per second (Mbps). It was recognized that both 2K chip sets and 4K chip sets would be used in projectors. The JPEG2000 compression permits a very high-quality 2K image to be displayed by a 2K chip set without having to decompress the 4K layer. This is an important and inherent characteristic of the wavelet compression approach used in JPEG2000. This means that the same image file is sent to all theaters, and whether a given screening room has a 2K projector or a 4K projector, the same file is able to be processed by all projectors as shown in Figure 6.5.

The picture track file is created by compressing the Digital Cinema Distribution Master (DCDM) described in the postproduction section of this book into JPEG2000 images at 250mbps, which generally results in about a 10:1 reduction in file size. Preprocessed forensic marking can be added if desired. The picture track file is typically specified to run at 24 frames per second (FPS). Within the standards, there are also provisions to specify playout at 25 FPS (sometimes used in Europe), 30 FPS (sometimes used for TV content). The Picture essence assets are most always broken into several track files labeled as "reels." The concept of reels stems from the days of actual film that had to be broken into many physical reels, each of which was roughly the same length containing 15–25 minutes of film, to be shipped to theaters. Today, the sizes of the reels vary widely, and are broken into segments that make sense to the production crew. Opening titles and end credits are often separate reels, and the main movie is typically broken into logical story segments. (Note that the segmentation may be chosen in anticipation of 'chapter' and menu structures that are often used in Blu-ray, DVD, and on-demand versions of the content.)

There is also a capability for picture branching. Using the previous example of an animated film with a stop sign that needs to be changed into a variety of languages, rather than replicating all the picture images in the whole movie, the DCP can include multiple copies of just the clip of frames that contains the stop sign, each copy having a different foreign language word for STOP. When the film is played out, the CPL instructs the projection system to branch out of the main picture, also known as the Original Version (OV) to the appropriate foreign version of the clip, and then branch back into the main picture. Again, this reduces the number of versions that need to be sent to all theaters in the world.

Stereo Picture

With the stability of the new digital projectors, coupled with advances in stereo viewing glasses, a resurgence of stereo presentations, or "3D movies," was seen starting in just a few years after the release of the DCI DCSS. While peaking in 2010 with the release of *Avatar*, stereo presentations have maintained a presence between 10% and 15% of the US box office, mostly on blockbusters, horror, and children's animated films.[7] While holding a small niche spot in theatrical presentations, stereo presentations have not garnered favor in home viewing. It is believed that wearing 3D glasses in a home environment have been a barrier for most consumers.

For theatrical stereo presentations, the picture track file will contain interleaved images for the two eye perspectives. Frame 1 for the left eye will be followed by frame 1 for the right eye, followed by frame 2 for the left eye, followed by frame 2 for the right eye, and so on. Stereo presentations are most commonly specified at 48 FPS (because the projector must present 24 FPS for the left eye and 24 FPS for the right eye simultaneously). Fifty FPS is used for some European stereo presentations and 60 FPS for some television presentations.

Stereo track files for a movie are not carried in the same DCP as the 2-dimensional (2D) picture files for a movie. The reasons for this are several. As film companies have improved their stereo processes, the 3D version of a movie often has slightly different editorial cuts than the 2D version of the movie, and "floating windows" trim off sides of the images to improve human perception of the stereography. Thus, audio, subtitles, etc. used in the standard 2D version of the movie, cannot be simply duplicated for the 3D movie version. Another complicating factor is that stereoscopic movies are often color-corrected at several different brightness levels, with 3.5–4 foot-Lamberts (fL) being the most common, and additionally at 7 fL and 10 fL for movies with larger distribution. The reason for this is that stereoscopic viewing systems typically result in the human eye receiving 50%–75% less light than is seen for a 2D movie and, therefore, there is a wide range of maximum brightness available at theaters throughout the world. To offer the best possible experience on all of these screens, stereoscopic movies are color-corrected at these multiple brightness levels, and each of these becomes a separate DCP. This creates an inventory, logistics, and distribution burden to ensure the correct version gets to the correct theater.

Standard Sound, Immersive Audio, Foreign Language Dubs, and Descriptive Audio

At least one audio track file must be included in a DCP. The original DCI DCSS identi-fied the use of both 5.1 and 7.1 surround mixes. Two channel, stereo presentations are not part of the specification. By far the most common format, the 5.1 surround mix, contains digital audio at 24-bit 48 kiloHertz (kHz), comprising six audio files. There are also provisions for a 7.1 surround mix at 24-bit 48 kHz, comprised of eight audio files, more commonly used on larger budget movies. Audio files are not compressed and are carried as linear Pulse Code Modulation (PCM) multichannel WAVeform (WAV) files.

In recent years, there have been several companies that have developed "immersive audio" for theaters, which has many loudspeakers positioned around and above the movie audience. For these mixes, up to 128 "slots" are allocated in an Immersive Audio Bitstream (IAB) for conveying any combination of channels and audio "objects." This IAB is carried in an Auxil-iary Data track file (see subsequent section on Auxiliary Data). The DCSS refers to the IAB as Object-Based Audio Essence (OBAE).

The transition from physical film to digital movies assembled by the projector allowed for many other advances, including the ability to carry many sound files. In the world of physical film, each movie theater in the world had to receive a copy of the film with the appropriate language audio track on the film. This required very careful inventory management and logistics management to ensure each theater had the right version. Because digitized audio is relatively small in comparison to the image files, a digital cinema package can include many dubbed audio files in many foreign languages. The actual file that is heard in the theater is determined by the CPL (see later section).

The transition from film to digital also provided easier access for people with some dis-abilities. A Hearing Impaired (HI) audio track is provided as a separate sound file, which boosts the dialogue and minimizes the music and sound effects to aid aging viewers with hearing loss. An audio description track is also included as a separate sound file for visually impaired (VI) audience members, which provides a narrated description of the action happening on the screen when there is no dialogue occurring. Consumers can hear these special tracks by using headsets available at the movie theater or by enabling accessibility applications on their Android or IOS devices. In late 2016, the United States (US) Justice Department issued a final ruling under title III of the Americans with Disabilities Act of 1990 (ADA) requiring all movie theaters to provide movie captioning (see next section) and audio description for all movies.[8]

Captions and Subtitles

Captions provide a written version of the spoken dialogue for viewers who are hearing impaired. Captions are occasionally offered as Open Captions, where the words are presented directly on the movie screen, particularly if a significant portion of the audience are hearing impaired. Closed Captions are more common, where consumers can see the captions presented on handsets available at the movie theater or by enabling accessibility applications on their Android or IOS devices. As noted above, this service was mandated to be available in all US movie theaters and was completely rolled out by June 2, 2018.

Unlike captions, which are in the same language as the dialogue, subtitles refer to a translation of the spoken dialogue into the native language of the viewing audience. These are displayed by the projector directly onto the screen along with the movie image. Subtitle tracks are extremely small files in comparison with the picture or the sound, so all subtitle tracks are routinely included in the DCP and the CPL determines which subtitle track is played with the movie.

Subtitle/Captions essence is generally structured as a timed-text .xml file that holds all the information about wording, size, color, location, and timing. One of two font style formats is needed; most commonly a TrueType Font (TTF) file defines the general font style to use for subtitles. Alternatively, and most often used for large character set languages, Portable Network Graphics (PNG) files can also be used.

Auxiliary Data

Auxiliary Data tracks were originally made available in the DCSS to help accommodate new technologies or processes not previously imagined, and are often used for such purposes on an on-going basis. As noted above, this has now become a standardized method for delivering immersive audio to theaters.

Other Unique Formats

While it would be most ideal if all the data for all movies worldwide could be carried in a single DCP, that is not practicable when it comes to multiple copies of the image file, which is very large. As a result, except for 2K and 4K resolution images – which are compressed into the same file – when different image files are required, they are most often placed into separate DCPs. There are several scenarios for which this occurs, and all of these create an inventory, logistics, and distribution burden.

There are provisions for standard 2D movies to be presented at 48 FPS, 50 FPS, 60 FPS, and 120 FPS for high frame rate (HFR) presentations. *Billy Lynn's Long Halftime Walk* was exhibited in 2D at 120 FPS in selected theaters. For stereo HFR presentations, the first three of these rates can be doubled to 96, 100, or 120 FPS respectively in order to present each eye with an HFR sequence of images. These files are compressed at 500 mbps (250 for each eye). *The Hobbit* trilogy was presented in 3D at 48 FPS for each eye (96 FPS) and *Billy Lynn's Long Halftime Walk* was presented at 60 FPS for each eye (120 FPS).

More recently, we are seeing the emergence of High Dynamic Range (HDR) content being presented in movie theaters. Because of the increased color volume available for these projection devices from the traditional movie projectors, a separate color-correction is created for these HDR-capable systems and is placed in a separate DCP.

There are some extraordinary presentations in some theaters that are not carried in the DCP, but instead are synchronized with the movie playout by using an output of SMPTE 12M Time Code from the projection system. Examples of this are some side screens, motion-based seats, lighting effects, and other atmospheric effects.

CREATING THE DIGITAL CINEMA PACKAGE (DCP)

MXF Wrapping and Encryption

Now moving to the turquoise boxes of Figure 6.5, the mastering track files are all wrapped using the Material eXchange Format (MXF). All image frames for a reel are wrapped into one video .mxf file. For a typical 20-minute reel of a 2D 24FPS movie, these files are often over 20 GigaBytes (GB) in size. All the audio channel WAV files (standard audio mix plus HI and VI) for a reel are wrapped into an audio .mxf file, creating a file that is about 1–2 GB in size. Likewise, the captions and subtitles are individually wrapped for each reel, resulting in files that are under 10 MegaBytes (MB) (PNGs being images will be larger than time-text). All total, most movies typically end up with a package in the range of 200–300 GB.

While encryption is voluntary, virtually all picture, audio, subtitles, and closed captions are encrypted using the 128-bit Advanced Encryption Standard (AES) using the Cipher Block Chaining (CBC) mode and using symmetric keys.

Composition Play List (CPL)

The Composition Play List (CPL) is uniquely created for each theatrical movie order and ties everything in the DCP together for a successful movie presentation. Well before a movie is scheduled to play, the Theater Management System (TMS) at a multiplex, or an individual projector system at a single theater, will read the CPL to determine whether all the specified components are available for the movie. The CPL contains a written human description of the movie in a field called the Content Title Text for use by theater managers/projectionists. Because early digital projection systems had limited space to display this field, the Digital Cinema Naming Convention (DCNC)[9] was developed by the International Society Digital Cinema Forum (ISDCF), which is still used today worldwide for the Content Title Text field. Later, at movie time, the CPL contains all the information needed for the movie playout. Just before playout, vital projector settings, such as whether the movie is 2D/3D, 2K/4K, flat/scope, FPS playout rate, 5.1/7.1 audio, etc. are provided to the projector. Then the CPL orchestrates all the appropriate essence assets together, turning what is otherwise just a random set of encrypted image, audio, subtitles, and auxiliary assets into a synchronized feature composition. The CPL breaks down the playout of the composition by reel. For each reel, the CPL defines which image .mxf, audio .mxf, and other optional essence .mxfs will be played out in synchrony. The projection system will read the CPL and pull all the specified tracks to render the presentation.

Asset Map, Volume Index, and Packing List

Once the essence mxfs and their orchestrating CPL are generated, the only thing left to do is to package those assets for delivery to a theater. This is where the Asset Map, Volume Index, and PacKing List (PKL) come into play.

All files within a DCP are referenced by a Unique Universal Identifier (UUID) rather than by file name. The Asset Map is the high-level control file type in the DCP, listing all assets within the DCP (expect for the volume index) and associating those assets UUIDs to their file path and name in relation to the location of the Asset Map.

Volume Index is a legacy file type used to break up a DCP over multiple storage volumes. When hard disk drives (HDD) were still expensive, this feature allowed breaking up a movie DCP over multiple storage devices, such as DVDs. The volume index was a simple tool used to track how many volumes (storage devices) were needed to complete the set.

Finally, with so much data contained within a DCP being transferred to theaters around the world, a method for describing what is in each delivery and for validating the content arrived intact was needed. The Packing list resolves this by listing all the files of a DCP (expect Asset Map and Volume Index) and associating them with a file hash. The Secure Hash Algorithm (SHA) 1 is used to ensure that a DCP arrived at the theater with no bit level file corruption errors. The Packing List (PKL) also describes the contents of the DCP sent in this transmission. It is possible, though not common, to send DCPs that are partially filled and to send additional components in a second shipment. As an example, this can be helpful if different track files are being produced in different parts of the world. It is also possible to send updated elements in a later shipment. For example, an updated reel of the end credits could be sent along with a new CPL without having to resend all the other picture reel tracks, audio tracks, subtitle tracks, etc.

KEY DELIVERY MESSAGE (KDM)

The Key Delivery Message (KDM) carries important security information and ensures the movie is only played on authorized equipment. A valid play "window" is provided, listing a start day/time and a stop day/time. Real-time watermarking, while optional, is almost always indicated as being required in the KDM. This security information, and all the symmetric keys that were used to encrypt the assets are then encrypted using asymmetric 2048-bit Rivest–Shamir–Adleman (RSA) public/private pairs to create a unique Key Delivery Message (KDM) for each projection system that has been authorized to play the movie. At playout time, the projection system must be authenticated in order for the KDM to release the encrypted files to be played. Because of this public/private key encryption technique, the KDMs can be transmitted to the theater by any means without worry of being confiscated or misused.

TRANSPORT

The big blue arrows shown in Figure 6.5 relate to the methods by which the DCP gets to the movie theater. From the introduction of digital cinema as a global distribution standard for theaters, physical delivery of DCPs has been the norm. Feature DCPs, generally ranging from 200 to 300 GB easily fit onto common, relatively inexpensive, 1 TeraByte (TB) hard drives. And while freight must be factored into the equation, physical delivery allows for quick and exact scaling to meet the month by month distribution needs for all motion picture studios, both small and large. Alternatively, for distribution of smaller packages like trailer DCPs,

USB thumb sticks are a common option. Independent of the physical media, storage technology, form factor, or connection specification; the industry has generally landed on using an EXTended (EXT) file system format of either EXT2 or EXT3. For mass duplication, methodology can vary, but often high-speed block-by-block duplication is used to copy content from a master drive to multiple slave drives at once. Validation methods can also vary, either using block scan to compare the master drive to slaves or utilizing the built in SHA1 checksums with a DCP to re-validate content after duplication.

Of course, with DCPs being a file-based format, electronic delivery is also an option and has become common across most regions of the world. Most electronic delivery networks consist of a digital transport path and edge device at the theater that receives and stores the DCP before handing it off to Theater Management System. In areas like North and South America where fewer localized versions are needed to cover large swaths of the population, satellites have been used as the most common digital transport path. Satellite distribution generally uses multicast protocols to send a single DCP to all satellite enabled edge devices in the network simultaneously, allowing for quick, cost effective, widespread distribution of content. However, as high-speed broadband infrastructure improves and with the potential of the Fifth Generation (5G) wireless networking just now rolling out, Internet-based delivery is starting to be more common. And with added benefits of faster targeted delivery and the ability to scale bandwidth to the needs of each theater, Internet-based delivery is likely to be the largest growth market for digital cinema distribution in the future.

THEATER STORAGE AND PROJECTION

Referring to the right side of Figure 6.5, initial digital cinema systems were generally a projector, with a computer server and storage devices alongside. With rapid advances in processor miniaturization, modern projectors now have slots to plug a server card directly inside the projector. With increases in storage density and I/O, some large theater complexes have centralized storage for their multiple projectors, which are controlled through a Theater Management System (TMS). And a TMS provides many other labor-saving capabilities for the theater, so the role of the TMS will be covered in more depth.

Theater Management Systems (TMS)

While DCPs can be directly ingested into a single projection system via a physical disk drive or other electronic delivery platform, most larger theater complexes employ a TMS to receive and verify all movie components are available, as well as to manage other aspects of the entire theater complex.

The TMS will directly ingest all the DCPs, CPLs, and KDMs and will identify when all components are available for playback, at which point operators can move components to appropriate auditoria. While moving components to localized storage for each projector is still the most common system arrangement, there are now systems available for multiplexes that have one centralized storage system servicing all auditorium projectors over a secure, high-speed network.

At its core, a Show PlayList (SPL) is an ordered list of what advertisements, trailers, and main feature will be played on a given projector. But an SPL often contains other common automated theater features. For example, auditorium lights may be automatically instructed to go dim lighting during advertisements and may then go to black when the feature starts. Other features, such as moving curtains, adding intermissions, having pre-/post-music, are also available. While creating an SPL may be done directly at the projector, it is more often accomplished using the TMS.

A TMS also provides a central location for theater managers to plan their theater schedules to maximize attendance. In the days of physical film, a projector would have one film platter mounted on one projector to be shown multiple times a day for a contracted number of weeks. Film platters were moved to smaller theaters as attendance for a given move lessened, but movement was not frequent, and rarely done during the course of a day. Further, a theater complex had a fixed number of film platters, so it was not possible to show the same movie in multiple theaters if there was more demand than was anticipated. With digital movies, it is easy to program a family movie for day hours and an adult movie in the evening hours, and it is easy to program that family movie in three theaters during the day, but only one theater in the evening. As a result, theater complexes are able to program their auditoria in response to their audience demands.

A TMS will gather the logs from all projectors, servers, and storage systems in the complex, thereby providing convenient status monitoring and report generation as well as allowing for easier and more-automated system maintenance.

Showtime Projection

Before playout, the projector will verify that it has appropriate keys and content for everything in its Show PlayList (SPL). At showtime, the projector steps sequentially through all the instructions in the SPL. At any step where a piece of content (advertisement, trailer, movie) is played, the KDM for that content must authenticate the projector before it will release the keys to decrypt the content. The projector will adjust its settings based upon the parameters provided in the CPL and then will proceed to playout that piece of content. When that piece of content completes, the projector will step to the next instruction in the SPL.

Projector Types

When building a movie theater, many aspects need to be considered in order to select the best combination of projector/screen/speakers/3Dviewing systems based upon the room architecture and key business purposes. A multiplex often has different products in different theaters in order to maximize the availability of the best viewing experience based upon the type of the movie being featured. It is beyond the scope of this book to discuss all of these parameters and the interactions among them. But because cinema projectors are in a significant period of change, they will be briefly described below.

When digital cinema started, all projectors were comparable to their film counterparts using xenon lamps as the light source. For stereoscopic (3D) presentations in very large auditoria, double-stacked projectors are sometimes needed in order to achieve the appropriate light level on the screen. The useful life of a xenon lamp for a theater projector is typically about 1,000–1,500 hours before its light output is cut in half and is no longer effective for projecting a movie on a theater screen. Further, xenon lamps are an expensive commodity requiring a hefty maintenance budget for theaters.

Much research has gone in to finding more efficient light sources. High-pressure mercury (HPM) lamps are similar in color volume to a xenon lamp, being just slightly lower in the red spectrum. Being less expensive and having a limited wattage output, these HPMs have made inroads for projectors in smaller auditoria.

In recent years, most projection manufacturers have brought a wide variety of laser-light-source projectors to the exhibition market, some best suited for small auditoria and some best suited for large auditoria. These products offer a much wider color volume as shown in Figure 6.6. The figure shows the full human visual color volume in daylight. Using red, green, and blue emission sources, the innermost white triangle represents the color gamut of the older International Telecommunications Union – Radiocommunications (ITU-R) Broadcast Television (BT) Recommendation Rec.709, which was designed to encompass the colors able to be replicated by Cathode Ray Tubes (CRTs). The center gray triangle is the traditional theatrical 3-point (P3) color gamut that a xenon lamp can create. The outer blue triangle represents a set of laser light source points specified in the modern ITU-R BT.2020. While requiring a higher initial capital investment, these Red/Green/Blue (RGB) laser projectors provide more colors and a consistent light output over a long period of time, thereby eliminating the expensive bulb-replacement maintenance budgets and resulting in a better overall return on the investment.

A drawback to projection systems is that the light is necessarily brighter in the center and falls off at the sides of the screen. This phenomenon is exacerbated by the use of reflective high-gain screens for stereoscopic 3D presentations. A potential future solution

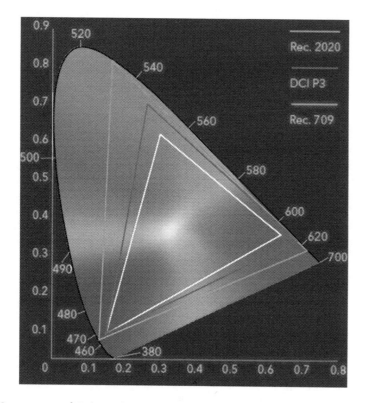

FIGURE 6.6 Comparison of Color Volumes[10]

is that of using a direct-view display, akin to those used in the home, which have uniform brightness across the entire display. These technologies, generally based on some form of LED, also provide a color volume closer to that specified in ITU-R Rec.2020. This notion is currently under development and investigation for future cinemas, but they are currently expensive and still have some technical issues of their own to overcome, particular in terms of blocking audio speakers that are positioned directly behind a movie theater screen.

SUMMARY – THE IMPACT OF DISTRIBUTION SYSTEMS ON PRODUCTION WORKFLOWS

As this chapter illustrates, television content distribution has evolved for over eight decades from its live linear analog broadcasting roots and the limitations of Cathode Ray Tube (CRT) displays to today's complex ecosystem of digital delivery of linear channels through broadcasting, cable, satellite and OTT systems, and pre-recorded content delivery in Discs, Cinema, and as on-demand content in cable, satellite, and OTT. While some progress was incremental, major technology disruptions took place with the advent of color and with the transformative shift to digital video technologies that was brought about by HDTV.

Most of the standards and video system parameters that are embedded in content production workflows are in one way or another a direct consequence of technology limitations and legacy compatibility considerations that have taken place in the context of evolving distribution systems. Examples include: 4:3 and 16:9 picture aspect ratios, 6 MHz analog transmission channels (and associated digital data rates), 59.94 Hz and 23.97 Hz fractional video frame rates, 24 Hz cinema frame rates, interlaced scanning formats, image resolutions related to 525 (480 active) vertical lines, rectangular pixels, YCrCb color difference signals with reduced bandwidth Cr and Cb, 4:2:2 and 4:2:0 sampling, "legal range" limits in video signals, "active and inactive" portions of video signals and the gamma nonlinearity.

While the vestiges of early analog distribution systems are just beginning to fade away, there are continually new legacy compatibility considerations being created that are inherent in the distribution choices made for HDR video systems, video and audio compression codecs, their associated resolution, video bit rate and bit-depth (e.g., 8 bit) limitations, storage size and data retrieval speed limitations, and the capabilities of digital consumer interfaces such as HDMI. The choices that are being made in today's most modern distribution systems will continue to impact content production workflows far into the future.

WRAPUP – CONTENT DELIVERY TO MULTIPLE DISTRIBUTION SYSTEMS

In this chapter we have separately discussed both television (broadcast, cable, satellite, OTT), pre-recorded media and cinema distribution to the viewer. While a great deal of content is created exclusively for television, it is important to understand that virtually all cinema content is also eventually distributed on television platforms. Feature films typically have "release windows" in which they are exclusively shown in theaters, then subsequently made available for purchase (on discs or by electronic sales), then subsequently available for rental on VOD systems and finally shown as a scheduled program in a linear television channel.

Because the viewing conditions and display technologies of a theater are so drastically different from those in the home television environment, digital cinema content and distribution formats cannot be used directly in television distribution systems. The ambient light level and the display brightness, dynamic range, and colorimetry are very different in the theater and the home. The perceptual characteristics of the Human Visual System (HVS) vary so significantly between the two environments that cinematic content must be separately color graded and transformed for television distribution. This frequently involves scene-by-scene creative decisions. This is also true for audio, where the acoustics and speaker distributions of a theater are different from the home environment, which is typically either 5.1 channel surround or stereo. And it is also true for captions and subtitles that need to be reduced in size to accommodate the smaller screen real estate.

With the advent of High Dynamic Range systems in both cinema and next-generation television systems, the number of content versions that must be created using additional color grading "trim passes" seems poised to increase. The situation is further compounded by the use of different HDR television systems by different distributors and in different regions of the world. Again, this is also true for audio, since next-generation television systems will include Immersive Audio as well as 7.4.2, 5.1 and stereo mixes.

And even as we consider next generation UHD and HDR television systems, we must remember that there is still a "long tail" of legacy HDTV and SDTV systems and pre-recorded discs in the consumer ecosystem, as well as analog television systems that still remain in service in many countries and in small cable systems in the US.

The basic elements of the workflows to transform cinematic content into the various television versions that are required for distribution to consumers are covered throughout the book. Such conversions typically include: color re-grading and conversions of color space, aspect ratio, image resolution, and frame rate. However, it is not just cinematic content that undergoes transformations for distribution to consumers. "TV" content frequently has worldwide audiences and is commonly transformed to different aspect ratios, resolutions, and frame rates for international distribution.

Once converted and represented in the correct baseband essence (video, audio, captions) formats, the video and audio compression and data formatting processing that are unique to each type of distribution system are successively applied downstream as content flows through the commercial supply chain. It is a common situation that content goes through several stages of compression and decompression on its path to the consumer.

NOTES

1. Advanced Television Systems Committee A/53 Digital Television Standard https://www.atsc.org/wp-content/uploads/2015/03/a_53-Part-1-6-2007-1.pdf.
2. https://www.dcimovies.com/.
3. https://www.statista.com/statistics/255355/number-of-cinema-screens-in-the-us-by-format/.
4. https://ieeexplore.ieee.org/browse/standards/number/smpte?queryText=%22d-cinema%22.

5. Reproduced by permission of Digital Cinema Initiatives, LLC (DCI). Copyright © 2005-2020 DCI.
6. https://jpeg.org/jpeg2000/.
7. https://techcrunch.com/2018/04/05/3d-movie-box-office-totals-take-another-dive/.
8. https://www.federalregister.gov/documents/2016/12/02/2016-28644/nondiscrimination-on-the-basis-of-disability-by-public-accommodations-movie-theaters-movie.
9. https://isdcf.com/dcnc/.
10. Courtesy of Christie Digital Systems: https://www.lamptolaser.com/fact7.html.

CHAPTER 7

Archive and Preservation

Section Editor: Brian Campanotti

INTRODUCTION

Once content has been created, it is critical that it be properly stored, archived, and preserved for future access, protection, distribution, and reuse. This may sound like a fairly straightforward task. However, when you consider the scale of data typically associated with media asset production, the many storage technologies and services available today and launching tomorrow, their associated costs, and the inherent need to protect these valuable digital assets for the long-term, you might start to appreciate the true complexity of the challenge.

Humankind has been dealing with the challenges surrounding storing, maintaining, and preserving information for many millennia. Whether we are talking about primitive paintings on cave walls or hieroglyphics etched into stone, it is clear humanity has strived to ensure longevity of its stories regardless of whether there was any long-term value, real or perceived. Even for the individual, ensuring aging film photographs and digital images will be accessible to our children and future generations can in itself present a very daunting challenge. How should they be stored? In what format? On what technology or cloud service? How should they be organized? How should they be described? These are the very same questions at the heart of the digital archive challenge.

With the exponential growth in data and with specific focus on the creation of content and opportunities for future monetization, distribution, repurposing, and even cultural reference, digital media archive and preservation is now and will continue to be an exciting and challenging topic.

In this chapter, we talk about libraries, storage, repositories, and, more generally, digital archives. Imperative to these *digital assets*, the associated metadata is paramount in finding, identifying, using, and reusing these assets again in the future. We will touch on considerations, technologies that can be leveraged, best practices, and how to maintain an archive to ensure the speedy and reliable retrieval of assets when needed now and into the future.

WHY DIGITAL ARCHIVE

There are various stages core to the creation of a media asset regardless of whether it is a simple personal project, short user-generated content targeting online streaming services, or a blockbuster production for a large media production company. Although simplistic, if we trace the lifecycle of a media asset from creation through to distribution/consumption, we can typically break it into several discrete, macro-stages, as shown in Figure 7.1:

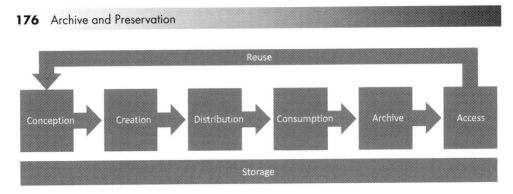

FIGURE 7.1 Media Asset Lifecycle

Although shown here as a sequence of steps, various pieces of this simplistic asset lifecycle often occur in parallel. In almost every case they all exist in some form or another. Interestingly, each and every one of these individual stages in the asset lifecycle inherently relies on discrete or shared storage but are typically only intended to be transactional in nature (i.e., online disk on an editing platform). The only stage which is not transactional or short-term in nature is the Archive stage which is where we will focus our attention, as it is the gateway to preservation, future access, reuse and repurposing, and monetization.

This asset lifecycle is obviously a large oversimplification of the complexities of the content creation ecosystem, but for argument's sake we can further classify this into yet higher-level, intrinsically interconnected macro-blocks as shown in Figure 7.2.

Creation

The first obvious stage is creation. This includes conception, acquisition, production, and other steps typical in the media creation process regardless of production scale. This encompasses everything from script and storyline creation through to the generation of a *final* consumable asset or set of assets. In many modern productions this creation phase could result in targeted products spanning movie theatre release, SVOD (Subscription Video On Demand) release, video and mobile gaming, VAMR (Virtual, Augmented and Mixed Reality), and an ever-increasing set of deliverables to attract, inspire, engage, and maintain viewers in an increasingly competitive and generally fickle consumer landscape.

Consumption

Consumption is the ability to reach end-viewers with the content generated during the creation phase. Again, this could be much more than a single, finished television episode or movie, as emerging platforms are capturing the attention of the viewer and the interest of content creators in a more meaningful and deep way to engage and display their creative art.

Archive

The final intrinsically linked stage is archive. This seemingly simple term has complex ramifications depending on your understanding and the scope and breadth of your implementation. These digital archives can be built to provide storage, protection, and preservation and even enable collaboration during the creative processes. All of these have specific implications which can be part of this stage and enable other stages in the asset lifecycle. Ideally, the objective is to implement a single digital archive system to cover all aspects of storage,

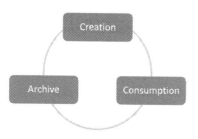

FIGURE 7.2 Simplified Asset Lifecycle

protection, and preservation whether we believe we need it or not. It is certainly easier and less costly to architect and build a digital archive platform for the future now than attempt to adapt one later on down the road out of necessity.

LIBRARIES, REPOSITORIES, AND ARCHIVES

You may often hear terms such as Digital Library, Digital Repository, and Digital Archive used interchangeably when referring to this final stage of the asset lifecycle, however they are quite different from a philosophical perspective and worth spending some time explaining.

Digital Library

> Library: A library is a collection of sources of information and similar resources, made accessible to a defined community for reference or borrowing. It provides physical or digital access to material, and may be a physical building or room, or a virtual space, or both.[1]

Although conceptually similar to the intent of digital storage and long-term preservation for digital assets, the likelihood of major studios and content producers allowing their content to be freely and openly accessed by the community is minimal. This would obviously conflict with the business models, which allows the creation of these valuable assets in the first place. However, this might very well be the case for historic and cultural archives, as dissemination of the information is the best way to ensure its continuity through the generations.

This topic is too complex to cover in depth here. Although very academic in nature, a large portion of the concepts, strategies, and practices developed inside this area of expertise are leveraged for the more general benefit of the digital archive.

Digital Repository and Digital Archive

These terms for a collection of digital assets stored in some fashion are typically used interchangeably and can be argued are valid in describing the concepts important to the industry. Effectively, the two terms Repository and Archive are defined as:

> Repository: A central location in which data is stored and managed.[2]
> Archive: A complete record of the data in part or all of a computer system, stored on an infrequently used medium.[3]

In fact, examining the definitions above, it is more accurate to describe our scope and objective as a Repository rather than an Archive, as the data we store will often share similar utilization patterns to more active storage devices used in other areas of the asset lifecycle.

To offer an accurate definition and hopefully lay out the scope of this chapter, we provide a somewhat enhanced definition below:

> Digital Archive: A central or distributed location where unstructured digital assets are stored on any type of storage technology in various topologies to facilitate long-term access, collaboration, sharing, repurposing, protection and preservation.

For simplicity we will use *digital archive* from here forward to refer to the concept of a central repository for digital assets.

PRESERVATION

Although clearly our objective is to store, protect, and *preserve* these digital assets now and into the future, we really must understand the significance of this daunting task. To help with really comprehending the scope and implications of the concept of preservation, we can first examine a definition.

> Preservation: the activity or process of keeping something valued alive, intact, or free from damage or decay.[4]

The interesting point to note is there is no clear timeframe given to this responsibility. The implication is this responsibility is a perpetual one theoretically lasting forever. Obviously, the scope of this statement can be vast as its implications are unbounded, especially when technology is concerned considering its rapid rate of change.

Fortunately, dedicated people embark on careers in preservation sciences to help ensure valuable assets are protected for the long-term, but we can hopefully learn and leverage significant points from this important area of study.

Library and Archive Science

Preservation is something that certainly falls within the scope of library and archive sciences but can and should play a part in all modern digital archive designs and implementations once the true scope is properly understood. To help understand its magnitude, let us examine the definition of Archival Science:

> Archival science, or archival studies, is the study and theory of building and curating archives, which are collections of recordings and data storage devices. To build and curate an archive, one must acquire and evaluate recorded materials, and be able to access them later. To this end, archival science seeks to improve methods for appraising, storing, preserving, and cataloging recorded materials. An archival record preserves data that is not intended to change. In order to be of value to society, archives must be trustworthy. Therefore, an archivist has a responsibility to authenticate archival

materials, such as historical documents, and to ensure their reliability, integrity, and usability. Archival records must be what they claim to be; accurately represent the activity they were created for; present a coherent picture through an array of content; and be in usable condition in an accessible location. An archive curator is called an archivist; the curation of an archive is called archive administration.[5]

Therefore, the scope of archive and preservation will outlive those responsible for these activities by definition. Not to imply that everything we create should be stored and protected forever but we can certainly appreciate the historical and monetary value of the original film print of Disney's Snow White and the Seven Dwarfs released in 1937. Do you think those involved in its creation could have contemplated having to properly maintain this asset when it was placed on a storage shelf close to a century ago? You can hopefully appreciate there is much more to a digital archive than simply storing and arranging files on a specific storage technology.

Further, in this age of deep fakes[6] there is ever-increasing importance in the curation and authenticity of these digital archives as trusted repositories and more than just simple archive storage.

> In library and archival science, digital preservation is a formal endeavor to ensure that digital information of continuing value remains accessible and usable. It involves planning, resource allocation, and application of preservation methods and technologies, and it combines policies, strategies and actions to ensure access to reformatted and "born-digital" content, regardless of the challenges of media failure and technological change. The goal of digital preservation is the accurate rendering of authenticated content over time.[7]

The essential guidelines for long-term digital preservation are outlined in the Open Archival Information Systems (OAIS) ISO 14721:2012 standard which has been a point of reference for the archivist community for many years:

> The information being maintained has been deemed to need "long term preservation", even if the OAIS itself is not permanent. "Long term" is long enough to be concerned with the impacts of changing technologies, including support for new media and data formats, or with a changing user community. "Long term" may extend indefinitely. In this reference model there is a particular focus on digital information, both as the primary forms of information held and as supporting information for both digitally and physically archived materials. Therefore, the model accommodates information that is inherently non-digital (e.g., a physical sample), but the modeling and preservation of such information is not addressed in detail. As strictly a conceptual framework, the OAIS model does not require the use of any particular computing platform, system environment, system design paradigm, system development methodology, database management system, database design paradigm, data definition language, command language, system interface, user interface, technology, or media for an archive to be compliant. Its aim is to set the standard for the activities that are involved in preserving a digital archive rather than the method for carrying out those activities.[8]

Essentially the OAIS reference model defines the scope of long-term preservation and guides one through all of the organization, process, and technical considerations key to achieving preservation goals and objectives. Although not often part of discussions in the entertainment industry, this is often the key reference manual for most memory institutions, libraries, galleries, and historical collection custodians and is certainly equally applicable to our industry.

The good news is there have been technological advances inside our industry where experts have incorporated many of the key tenants of the OAIS Reference Model into systems which store, protect, and preserve digital assets. One example is the Archive eXchange Format (AXF) standard covered in more detail later in this chapter.

There are obviously entire university degree programs designed to prepare people for careers in library and archive sciences, so the full study and its implications are far outside the scope of this chapter. That being said, we will touch on some of the core concepts throughout in helping to build a clear and concise case for the digital archive.

Forever Is a Long Time

In the scope of preservation as it applies to digital archives, we note that we must consider an infinite timeline for our activities to protect and ensure future access to these valuable assets. Of course, it is unrealistic to plan the implementation of a digital archive for eternity, as the cost would likely present an insurmountable barrier, making the initiative impossible to justify from a purely business perspective. That being said, we also cannot effectively plan a digital archive around a finite timeline with the expectation something magical and transformative will happen to alleviate our burden in the future. Unless there is a firm belief the content being produced is simply not worth protecting, which is certainly impossible to judge looking into the future.

We can be pragmatic about this point and ensure we are planning for at least the next decade and make good architectural and sound technical decisions to ensure a higher level of confidence in the sustainability of the digital archive for this more modest *long-term* time horizon.

WHAT IS AN ASSET

We have been discussing the storage of assets in our digital archive, but perhaps it is worth taking an intentional step backward to fully comprehend what an asset represents as it applies to our objectives around storage, preservation, and accessibility for the long-term:

> As·set: a person or thing that is valuable or useful to someone or something.[9]

Asset Value

The value or perceived value of an asset can be measured in many different ways, including monetary, historical, cultural, or any combination of these.

The first obvious one is monetary and is normally the motivation for commercial content creators to contemplate any long-term storage and protection strategies. There are others which are arguably much more significant as we are often delving into historical collections to recall or derive references for points in time or tracing back cultural or eco-political data from the past periods. News production is an obvious example where current stories are often enhanced by leveraging historical footage from past events adding context, depth, and even proof of accuracy. Even for major commercial productions, research teams will often scour period footage from various historical archives to capture cultural references and storyline context and even align historical facts from the era.

In addition to the obvious commercial motivations, many important organizations such as the United Nations,[10] UNESCO,[11] and the Shoah Foundation[12] have embarked on digital archive projects to protect and preserve soon-to-be-forgotten cultures, languages, and significant historic events. They are building digital archives containing interviews and first-person testimonials while witnesses to these significant events are still alive. The intent is to ensure these cultures, languages, and significant historic events can never be rewritten, reshaped, or forgotten while providing unmatched archive collections to researchers, scientists, and future generations.

Fundamentally, an expectation of *value* can be considered a prerequisite in the creation of an asset and certainly the key motivation to store and protect it.

The Asset Construct

An Asset comprises two fundamental components: essence and metadata.

In our case, essence is the video, audio, image, or other data which comprises the actual content of the asset. Metadata is information about the essence such as filename, description, transcripts, context, size, duration, and other critical information. Both of these come together to ultimately define the usefulness and value of an asset and are critical to the true value of a digital archive. Most organizations do a fine job of storing essence but are not always good at storing an asset.

Imagine having only a list of simple files (i.e., AB12.xyz, CD34.xyz, etc.) stored on a storage device. As long as the person who originally created these files or someone who they shared the information with is available, context can be discovered adding meaningfulness to these files. However, we only need to scale this challenge to a few hundred or thousand files to demonstrate the impossible challenge of recalling important relationships, context, and details. We cannot simply rely on an *arms-length* relationship between these files and the metadata which describes them.

To better understand this asset concept, let's look at an example we can all easily relate to of a digital photo of a landscape taken on our mobile phone. If someone views this image with nothing to describe it or add context to it, they may still enjoy viewing the photo but have little or no context or understanding of the image or its potential significance. Further, they may even misinterpret or misrepresent it to others. However, if the mobile phone assigned a name to the image with a date and timestamp we already have some additional comprehension of the asset, adding to its value. If the photographer also adds a simple note to the image indicating it was taken at Algonquin Park, Ontario, Canada during a family vacation, the viewer can understand even more about the image and its context, essentially adding to its perceived value and authenticity. If the mobile phone automatically augments the photograph with GPS metadata, the viewer can now locate exactly where the photo was taken on a map, adding even more depth of understanding and perhaps even a greater appreciation of the image. This valuable *metadata* can then be used to cross reference additional metadata from external services such as finding similar images or examining weather data for exact environmental conditions at the time and location where the photo was taken.

As you can see, the picture, or *essence*, itself is only a portion of the true value of the asset which is inherently dependent on an immutable connection with its associated *metadata*. This is a core requirement for digital archives as essence may be valuable only as long as the people who created it are still around to assist with finding and comprehending it. Therefore, the essence is not worth archiving or preserving unless there is associated metadata immutably connected to the essence, forming a valuable asset worthy of storing in a digital archive. If we imagine this same exact digital photo example "at scale" across thousands of digital images in our individual photo collections, you can easily start to realize the relatively insurmountable challenges facing content creators, broadcasters, and custodians when architecting and constructing a digital archive where metadata is not present.

Even a fairly simple and rudimentary folder naming convention (i.e., "Algonquin Park Family Vacation – June 19, 2020") adds significant value to the essence and helps form an asset. This simple metadata allows us to quickly search, locate, and deduce useful information about the files contained in this named folder. The point here is even the most basic metadata can be extremely valuable in the ability to find, access, and utilize these digital assets now and into the future. We should be careful to ensure metadata is in place as we plan our digital archive but not get blocked looking for metadata panacea since it rarely exists.

The scale of this metadata challenge for broadcasters, content creators, and historical curators with thousands or even millions of assets can be crippling. Without key metadata associated with each essence file, these valuable collections become inaccessible and ultimately are at risk of being lost as they simply cannot be effectively searched.

Recognizing the importance of metadata, many organizations are starting to implement strict metadata collection and curation processes for newly created assets throughout the entire asset lifecycle, right from the conception phase. For legacy asset collections where limited information is available, Artificial Intelligence and Machine Learning (AI/ML) tools are now enabling automated metadata discovery and metadata enrichment for large-scale audio/visual archive collections.[13] These AI/ML tools can identify and catalog human faces, animals, locations, objects, landmarks, and more, all while transcribing multi-language audio tracks rendering them searchable. These tools can be used to automatically augment audio/visual essence with searchable metadata automating these historically challenging metadata tagging tasks at scale. These algorithms can learn from the collections themselves and be applied iteratively to continually enhance and deepen our comprehension, and increase the resultant value, of these digital assets.

As AI/ML technology continues to advance, they will likely be the salvation for aging archives and likely bring valuable historical and cultural assets back to the forefront for future generations.

Technical Metadata

On the technical front, knowing what image processor or compression algorithm was used for our photograph, video, or audio recording will give us higher probability of being able to access them (read, view, listen, etc.) in the future. Of course, we can use rudimentary mechanisms such as filename extensions to give us hints, but the deeper and broader this technical metadata is, the more likely we are to benefit from it later.

The good news is most modern systems automatically generate very deep and accurate technical metadata for digital assets including compression details, frame size, frame rate, timecode values, duration, etc. Most also include automated contextual metadata such as GPS data in combination with asset provenance. Provenance is from whence the asset came and can include the username, camera, device, editing platform, or similar device which created the asset along with date, time, etc. This helps future generations to comprehend the birth and life of the asset, an ultimate key to true preservation.

If each time we modify or enhance the digital asset throughout its life, we are careful to include information from where it came and who is currently acting on the asset, we can help future generations better understand an asset they are examining and ensure its authenticity, another important tenant of our digital archive objectives.

Connecting Metadata and Essence

If we remember back to the days of film cameras and printed photographs, we often scribbled a couple of words or the date (its *metadata*) on the back of the photograph so we could later understand the context when viewing it. When we write metadata on the back of the printed photograph, we have fundamentally and immutably connected the essence and the metadata, and therefore formed an asset which now has inherent long-term value worth preserving.

When designing our digital archive, we need to ensure there is an immutable, pervasive, and sustainable connection between the asset essence and the metadata for the long-term. Historically, the metadata was stored and maintained in systems separate and distinct from the essence storage adding to the complexity and risk regarding future accessibility and preservation. Often times, metadata is stored and maintained in Media Asset Management (MAM) systems or other external database, while the essence is stored on file-based storage devices. There exists a fundamental link or key between these systems which connects the metadata to the essence, but only the combination of both systems can be considered a digital archive storing assets. Obviously, there is an increased risk here when systems are upgraded or replaced where these loose connections between essence and metadata could be lost.

There has been a lot of work in the area of fundamentally connecting essence with metadata to help overcome these potential challenges in the future. On the technology front, many modern object and cloud storage systems allow key/value metadata tagging of assets when created, allowing a fundamental and transportable connection between the two. Standards have also been developed to immutably merge essence and metadata repositories into a scalable asset platform which offers a more sustainable and risk-free digital archive future and will be discussed later in this chapter.

In addition to the access, search and retrieval benefits of fundamentally connected metadata, the capability to dynamically and intelligently orchestrate digital archive storage based on asset metadata is an exciting emerging field. Most legacy digital archive systems rely only on policies built on rudimentary file parameters such as size, age, filename or other static parameter to unintelligently govern storage orchestration. This metadata-driven orchestration approach is further enhanced when considering the metadata enrichment and metadata mining capabilities of modern AI/ML tools. This essentially adds self-evolving intelligence to digital archive orchestration based on formerly unknown or undiscovered information about the assets themselves. More on this exciting concept later in the chapter.

DATA SCALE

As our industry continues to generate an increasing number of assets, across broader plat-forms and delivery mechanisms and at increasing resolutions, the scale of data is increasing at an exponential rate.[14]

The following table gives modest examples of the amount of data one single hour of content represents at common resolutions and framerates in use today. Despite the large numbers, this is only a modest view of the actual data scale, as oftentimes the entire asset lifecycle can result in 10× to 100× more data throughout the process than is represented in the table below.

Format	Nominal Bitrate (Mbps)	Nominal Bitrate (MBps)	Frame Rate (fps)	Frame Size (MB)	One Hour of Content (GB)
SD	50	6	30	0.2	22
HD	100	13	30	0.4	44
Uncompressed SD	270	34	30	1.1	119
XAVC 4k	330	41	24	1.7	145
ProRes422 HQ 4k	754	94	24	3.9	331
ProRes4444 4k	1100	138	24	5.7	483
BlackMagic4k	1400	175	24	7.3	615
Uncompressed HD	1500	188	24	7.8	659
Sony F65 RAW (3:1) 4k	2000	250	24	10.4	879
ProRes4444 XQ 5k	2100	263	24	10.9	923
Canon RAW 12bit 4k	2300	288	24	12.0	1,011
Uncompressed 4k	12000	1,500	24	62.5	5,273

FIGURE 7.3 Data Scale in Media and Entertainment

If we consider a major content production company, studio, or even independent pro-ducer with thousands or millions of hours of content, the scale of data which needs to be considered when building a long-term storage, protection, and preservation strategy is daunt-ing. Key to this challenge is taking some time to contemplate and understand your ultimate objectives early on in the creation process. If you plan to "keep everything forever" then you need to attempt to calculate the final amount of data, ensure budgets capture this objective, set enough money aside during crunch time to ensure you can meet your objectives, and have a solid understanding of what "forever" actually means. Oftentimes, what begins with noble objectives becomes a burden later on when the production is completed, and some-one is tasked with maintaining an unsustainable objective potentially placing these valuable assets at risk of being lost.

As an interesting side note, this is not a new challenge driven by digital production or the data volumes identified above. This has always been a challenge dating back to the days of film and video tape. Placing a recording of your production onto a shelf only solves the "forever" problem for a decade or two even if stored in ideal environmental conditions. Recording media ages and player devices fall out of production and then out of support. You can have all of your assets fully protected and stored, but if you cannot play or recover the data contained on them, then you may as well have destroyed the media earlier on in the process. This leads into the need for periodic maintenance of these media storage technolo-gies to ensure continued access, on sustainable media, a topic we will cover in more depth later in this chapter.

ENTERPRISE AND BROADCAST TECHNOLOGY COLLIDE

Continuing on our path of setting the stage for the digital archive, it is likely important to differentiate terms often mistakenly used interchangeably and introduce a modern take on the concept of the digital archive.

Backup

These solutions made their way into our industry based on their widespread use in general enterprise Information Technology (IT) environments. Although seemingly in line with our objectives outlined so far, IT backup technologies are fundamentally contrary to the philosophical directives behind the digital archive.

File backups are made to protect critical data but with the underlying intent of never having to access this information if primary systems are healthy. The typical use-case involves data being *backed up* on a regular basis with the hope it will never have to be accessed. Backup systems can also be used to support disaster recovery by allowing off-site copies of user data to be maintained in the case of a catastrophic event (i.e., fire, flood, ransomware attack). In the case of data loss, these backups can be accessed and affected data safely recovered. This is often a very manually intensive operation but offers protection against data loss.

However, when examining our objectives in building a digital archive, our key motivators in addition to storage were access and reuse which backup systems are fundamentally not designed to handle. Although they often appear the same as digital archives on the surface, they are tuned to be very good at storing but not as great at regular, random data access and reuse. Further, these IT backup systems are always *file-based* implementations and handle the asset essence only. This necessitates metadata be maintained on the original systems which generated the files further reducing their applicability to our digital archive objectives.

Hierarchical Storage Management (HSM)

Similar to Backups, HSM solutions also made their way to the entertainment industry via enterprise IT. Although seemingly similar in appearance, they also embody concepts fundamentally contrary to the philosophical directives behind the digital archive.

HSM systems assume file-based data which has not been accessed in some time is *less important* than data recently created or accessed. HSM systems use this fundamental concept to *tier* (move) this less important, aging data to less expensive storage. This assumes the data will be used less and less as it continues to age and therefore can be stored on less accessible and less expensive storage technology. HSM workflows in our industry may be valid for the tiering of data to less expensive media, but the demands tend to be very unpredictable causing performance bottlenecks in the handling of important retrieval operations from these less accessible storage tiers.

Typical HSM policies may instruct the software to move files not accessed in more than 30 days off primary storage to lesser speed and therefore lower cost storage, with the eventual target typically being data tape. HSM solutions use the clever concept of *stub files*[15] to make user-data appear as if it still resides on primary storage, but allowing it to be moved to cheaper storage tiers leaving just a remnant, or stub, of the original data behind. When users attempt to access data which has been moved by the HSM system, they are simply instructed to wait while their data is retrieved from slower storage tiers. As the usage patterns of data in traditional IT environments tend to support the underlying concept of data aging, HSM systems are frequently deployed to help reduce overall storage costs with minimal user impact.

Additionally, HSM solutions are *file-based* by their very design and cannot address the complex challenges in these media-centric, big data environments. This simple policy-based file movement between storage tiers does not take into account specific files are often components of more complex assets, and not the assets themselves. As file migration policies are satisfied, the missing context of the object can cause components of a single asset to be scattered between storage layers and amongst several data tapes in the data tape storage tier. This limitation is further exacerbated in unstructured environments where the inability to predict the nature and content of assets make adequate file-handling policy definition impossible limiting agility and adding management challenges to the environment.

Several vendors have worked to adapt the core features of HSM to better fit the digital archive applications in our industry, but they always suffer these same fundamental limitations. Ultimately, HSM solutions are file-based as was the case with backup solutions discussed previously, and their applicability to our digital archive objectives are similarly limited.

Content Storage Management (CSM) and Active Archive

CSM and Active Archives are very similar concept and are designed specifically to serve the needs of digital archive applications in the media industry.

The term Content Storage Management[16] was coined in 2006 by Brian Campanotti and Rino Petricola (Front Porch Digital) to describe these differentiated archives specifically targeted toward object-centric, large-scale unstructured data sets and high-demanding usage patterns typical in digital archive applications. These systems offer benefits as they deal with essence files as *objects* rather than just individual files and handle these complex relationships natively. The term Active Archive was coined in 2010 by members of the Active Archive Alliance[17] to describe a similar digital archive system approach.

Although more attuned to the specific needs of the industry, they both fall short in our pursuit of a modern digital archive. These solutions continue to handle storage

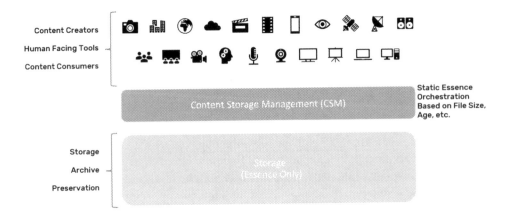

FIGURE 7.4 Content Storage Management

orchestration at the essence level, albeit at the object essence level, essentially missing the opportunity to intelligently and adaptively orchestrate assets leveraging dynamic metadata.

Many legacy CSM vendors are struggling to adapt their solutions to better handle the modern demands and challenges of digital archiving. However, layering on modules in an attempt to add modern intelligence to aging CSM platforms is similar to what the HSM vendors did in an attempt adapt their solutions to the media industry in the early days. There is an emerging opportunity to build a modern digital archive solution to better handle not only media assets based on their essence characteristics at the object level, but also to handle these complex, massive-scale storage orchestration challenges leveraging dynamic asset metadata as well.

Software Defined Archive (SDA)

The Software Defined Archive (SDA) concept was first introduced in 2017 by Cloudfirst.io[18] as a modern approach to the digital archive. Valuing the past decade or more of digital archive technology evolution, the SDA embraces all of the fundamental capabilities of an Active Archive or CSM solution (i.e. treating essence file collections as objects, etc.), but possesses the ground-up capability to dynamically leverage asset metadata to build a cognitive, metadata orchestrated digital archive system.

Although similar in architecture to CSM solutions, this SDA approach enables the orchestration of media assets on more than just static polices based on its constituent files or objects (i.e. file create date, size, age), handling them as *assets (metadata + essence)* rather than just the essence files which comprise them. The SDA essentially constructs an intrinsically connected essence and metadata repository which can each be individually enriched, orchestrated, and leveraged to best handle modern digital archive workflows in a fully automated fashion. With the addition of AI/ML metadata enrichment, asset metadata can continuously improve adding significant depth to the capabilities of the digital archive system, enabling it to learn and self-adapt over time.

Rather than defining static policies which loosely map to service level agreement (SLA[19]) targets, these dynamic systems can automatically adapt to different requirements based on

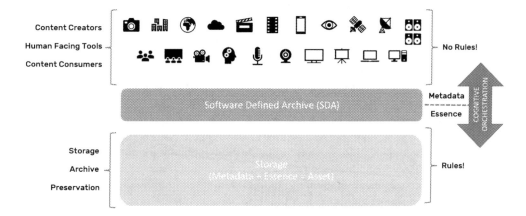

FIGURE 7.5 Software Defined Archive

rich and deep asset metadata. Imagine having a digital archive that dynamically adapts to trending news stories or sporting event schedules by bringing the most relevant content to more accessible storage automatically, alleviating any requirement for manual intervention. Conversely, the SDA can dynamically route assets to the most cost-effective tier based on current market storage rates while ensuring the most valuable content is maintained in multiple geographic locations and on divergent storage technologies for the highest level of protection possible.

As a simple example, the SLA for a specific production asset may require two copies in the cloud, another may require one copy on the least expensive storage available and another may demand two copies across different cloud providers and one on-premise. The SDA will automatically orchestrate on an asset-by-asset basis to achieve these SLA targets based on individual asset metadata (i.e. production group, inventory number, series, episode, etc.). Further, as SLA targets or asset metadata evolves over time, the SDA always self-reconciles to ensure assets adhere to their specific SLA requirements alleviating much of the administrative burdens typical in legacy archive environments today.

Unfortunately, legacy CSM and Active Archive systems are not able to embrace these modern approaches to true digital asset archiving as they are fundamentally built with a *storage-centric* focus rather than the necessary *asset-centric* focus. In many cases where these legacy archive systems are in place, the necessary *archive transformation* may appear to represent a nearly insurmountable challenge. Archive modernization should be carefully considered, examining factors such as cost, return on investment, workflow benefits, and other key factors, all while ensuring you are simply not delaying an inevitable migration off an aging legacy archive system. Archive transformation will be covered in more detail later in this chapter.

Build versus Buy

This has been a perpetual challenge throughout time. Do software and hardware vendors intimately understand an organization's operation intimately enough to build a suitable digital archive? Will they meet our specific needs? Is the vendor equally invested in our long-term digital archive objectives?

Arguably, digital archive software development is easier now than ever before with the proliferation of open-source components and standardized interfaces available to most modern storage technologies (i.e. S3). However, experience and countless case-studies clearly demonstrate that open-source is far from free and burdened with the risk of long-term unsustainability.

Digital archive vendors invest significant time and money in architecting and developing solutions to best serve the broader needs of these complex digital archive environments. Although they often cannot be customized to specifically fit any one particular environment, they are best at serving the general objectives, leveraging a broader and rich view of the industry as a whole. This potentially brings additional experience and depth to each new deployment in being able to leverage requirements and features developed across dozens or even hundreds of other forward-thinking organizations.

Custom built applications tend to be the best fit early on but become very difficult to continue to sustain and advance once attention (and budgets) turns to focus on other challenges facing the organization.

Instead of embarking on an in-house digital archive development effort, organizations are better served focusing on assembling a vendor-agnostic commercial engagement and selection of technologies which protect against long-term asset lock-in. As long as the digital archive and/or cloud solutions being considered allow configurable, metadata-centric orchestration policies to be defined and maintained without involving custom development and introduce no proprietary interfaces, asset wrappers or containers, this may be sufficient to ensure long-term asset protection.

There have been several cases over time where a digital archive or storage vendor decided to change their focus or even abandon their products altogether leaving existing users orphaned and desperate to free their assets. Risks need to be considered on both sides of this equation.

The AXF standard covered later in this chapter goes a long way to protecting and preserving your most valuable digital assets and helping avoid vendor lock-in. In fact, through a series of recent events, this hypothesis was proven true as several digital archive vendors were able to natively support the migration of data archives created by other vendors solutions eliminating the necessity to migrate the entire collection prior to decommissioning the legacy vendors solution.

STORAGE TECHNOLOGY

Fundamental to every backup, HSM, CSM, SDA, or more generally, the digital archive environment is storage. Storage evolves rapidly trying to keep pace with the ever-increasing demands for capacity and speed while optimizing short- and long-term costs. Modern digital assets (like the photos on your mobile phone) will often touch multiple types of storage throughout its lifecycle as is the case with media assets in these digital archive environments. At scale, there are no one-size-fits-all approach, and often these digital archive systems will employ various types of storage, technologies, and generations to achieve their ultimate goals which will also evolve significantly over time. Obviously, the velocity of change and

Technology

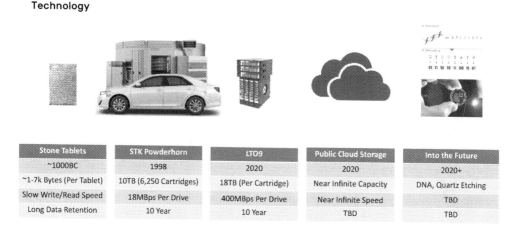

Stone Tablets	STK Powderhorn	LTO9	Public Cloud Storage	Into the Future
~1000BC	1998	2020	2020	2020+
~1-7k Bytes (Per Tablet)	10TB (6,250 Cartridges)	18TB (Per Cartridge)	Near Infinite Capacity	DNA, Quartz Etching
Slow Write/Read Speed	18MBps Per Drive	400MBps Per Drive	Near Infinite Speed	TBD
Long Data Retention	10 Year	10 Year	TBD	TBD

FIGURE 7.6 The Evolution of Archive Storage Technology

advancement in storage (and compute) technology makes any level of depth here quickly irrelevant but is included for context to the overall discussion surrounding digital archives. Figure 7.6 shows the abbreviated evolution of digital archive storage technologies through the recent millennia.

The evolution of digital archive storage over the past two decades has taken us from a robotic device larger than the average family car to a single data tape cartridge with close to twice its capacity, enabling more than 11,000× the storage in the same physical footprint. One can only imagine what the next 20 years will bring!

Despite these impressive advancements, the industry is certainly in need of a storage revolution. Essentially, we are still using the same fundamental technologies we have been for the past 50 years (i.e. transistors, magnetic tape, magnetic spinning disk platters), simply modernized, more dense, more reliable, and lower cost but without any substantive disruption.

Unfortunately, without a significant value proposition to the greater IT/enterprise market, no one can effectively, sustainably, and profitably build technologies specific to the entertainment marketplace to meet its unique demands for cost-effective capacity. Packaging these broader technologies into devices and appliances with unique software which resonate in the entertainment market is normal, but real storage innovation is difficult without the market opportunity to drive it. Ultimately, we will continue to be a slave to storage innovation in other verticals and rely on software and applications to make it relevant to the entertainment market. The good news is, there are current and emerging options which can be effectively combined to meet the specific demands, both technical and commercial, of these digital archive environments.

The following sections highlight storage technologies commonly used in the construction of digital archive systems today.

Solid State

Solid state or flash storage refers broadly to any storage technology which involves silicon-based, persistent memory devices. These contain no moving parts and are typically faster and more reliable than mechanical devices which rely on magnetically aligned particles to store data such as hard drives and data tapes. Solid-state drives (SSDs) are the most common and are usually comprised of various types of solid-state memory. Newer Non-Volatile Memory Express (NVMe) devices demonstrate much higher performance but also carry a higher cost and are therefore more specialized in their applications.

Flash storage is common inside consumer devices such as mobile phones, media players, and smart watches and recently more pervasive in modern laptop computers due to decreasing costs and positive impact on battery performance.

Hard Drives

Hard drives, or more generally, magnetic disks are by far the most pervasive storage devices used today across all industries and applications. High-speed and high-capacity hard drives can be mixed in a single storage device to provide high performance, high capacity, or a balanced combination of both. Many hard drive technologies are in existence with more on the near-term horizon.

Serial Attached SCSI (SAS) and Serial Advanced Technology Attachment (SATA) drives are typically used in higher performance environments. Shingled Magnetic Recording (SMR), Heat-Assisted Magnetic Recording (HAMR), and Microwave-Assisted Magnetic Recording (MAMR) drives are typically larger in capacity and less expensive and therefore targeted at scale-out storage solutions.

Various hard drive technologies offer unique performance and cost profiles and are best suited to specific tiers of storage based on these characteristics. Most often individual hard drives are combined into an array of disks (i.e., RAID). This increases aggregate capacities and improves performance by striping data across multiple drives while offering increased resiliency with the addition of parity data. Individual hard drives and disk array controllers often incorporate solid state cache to improve transactional and read performance over that of the disk alone slightly blurring the boundaries between these discreet storage device types.

Hard drives continue to earn their place in most storage environments, but many ominous technological challenges such as ariel density, superparamagnetism, and others must be overcome to ensure their relevance in the coming years. There is a need to dramatically increase capacity and performance while reducing costs to continue to make them compelling compared to solid-state drive technologies which continue to decline in price due to wide-scale deployment.

Optical Disc

Optical disc has been fighting for relevance in the storage world for many years, held back due to limited adoption, slower performance, and higher costs. The only optical storage technology still commercially available is Blu-ray disc from both Sony and Panasonic. A handful of deployments of optical storage as a digital archive storage tier exist today, primarily as it offers a technologically diverse approach with copies maintained on optical media in addition to traditional data tape or other magnetic media.

Current generation optical discs are significantly less dense and offer much slower performance as compared to competing data tape technologies and have yet to find a critical value proposition to motivate their wide-scale adoption. It is unclear whether the current generation of optical discs will continue to be relevant in these complex and growing digital archive environments without significant near-term advancement.

There have been many companies working on commercializing novel optical technologies such as holographic and others, but none have had a compelling enough case against the continued advancement of mainstream storage technologies such as flash, disk, and data tape.

Data Tape

By far the most pervasive, large-scale storage technology is data tape. When the entertainment market shifted to go tapeless (video tapeless) more than a decade ago, many companies building large-scale digital archives began relying more heavily on data tape for storage. Obviously, data tape does not have any of the legacy issues that video tape had, but it is still based on the very same principles. Of course, it is denser, faster, more reliable, and less expensive than ever before, and these advancements continue generation after generation ensuring its continued near-term relevance for digital archives. Cost-wise it currently remains the least

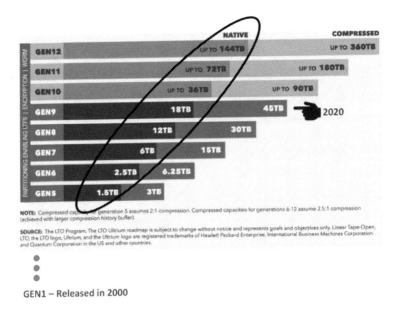

FIGURE 7.7 Evolution of Data Tape (from the LTO Consortium at lto.org)

expensive raw storage technology available on the market and is therefore very well suited for on-premise digital archive storage.

Figure 7.7 shows the generational evolution of the most common data tape technology available today, Linear Tape Open (LTO). LTO has continued its generational advancement for more than two decades now since the launch of LTO-1 in 2000 and is by far the most deployed data tape technology in the industry.

These data tape drives and data tapes are almost always deployed in scalable robotic systems with tape slots which can physically store thousands of individual tape cartridges, enabling massive-scale storage footprints. Aggregate tape robotic system capacities are into the exabyte (EB) range with the ability to read and write terabytes (TB) of data per hour leveraging current data tape technology.

Robotic picker arm(s) are used to quickly and reliably move tapes from their home slot (called bins) to a free tape drive where the particular read or write operation can be serviced by the archive software. Once the request has been serviced, the robotic arm(s) moves the tape back to the home bin and then awaits the next request. Tapes can typically be loaded/unloaded in a matter of a few seconds, and tape libraries can contain up to eight robotic arms all working in parallel for massive-scale physical operations with libraries containing hundreds of individual data tape drives. In addition, these tape libraries offer the ability to import/export tapes to support offsite duplicates and limitless storage expansion with external shelving units to house data tapes. While this physical shuffling of data tapes may seem archaic, many businesses are moving back to this type of technology to provide increased protection against cyber and ransomware attacks levering the *air-gap* protection provided by data tapes sitting on tape shelves physically separated from the data tape drive that can read and write them.

Unfortunately, organizations often have a love–hate relationship with data tape.

The most disliked characteristic of data tape is the need for generational migrations which are very expensive and time consuming at scale. This cycle typically happens every eight to ten years as tape technology ages and new generations are released. Each new generation is faster, denser, and typically more cost-effective per capacity unit. As the digital archive data footprint grows in size, these systems must be periodically upgraded to ensure they are ready for the ever-increasing onslaught of data. Also, the risk of asset loss due to aging and unsupported or even unavailable tape drive technologies is a certain motivator for regular system migrations.

In fact, this migration challenge is not a problem exclusive to data tape as all storage technologies must also be migrated and replaced every so often as they age, fall out of support, or simply begin to fail. This is often referred to as a *forklift-upgrade* as it requires the physical swapping of infrastructure. When migrating a disk array to another disk array, the process is typically relatively quick due to the performance of disk and the fact these arrays are several orders of magnitude smaller in capacity as compared to large-scale tape archive systems.

With data tape library migrations, each individual tape must be loaded into a tape drive, its data read to a cache location (typically disk drives on an external server), and subsequently written back to a new data tape. Each mount, read, write, and unmount cycle can take several hours, as each data tape can store vast amounts of data. This process then repeats hundreds or even tens of thousands of times to properly migrate all tapes in the digital archive. The mechanical characteristics of tape combined with the scale of these library systems often reaching into the dozens of petabytes (PB) range, migration can be a long process, often taking months or even years. The good news is most digital archive software can handle this as an automated background task alleviating the management burden on the system administrators but still requiring a significant investment of time and money.

These data tape robotic systems also occupy large physical footprints inside customer data centers and require significant maintenance and upkeep as they are large and complex mechanical devices. With a general industry trend toward facility consolidation and away from customers building and maintaining their own data centers, deploying and maintaining these large data tape library footprints is becoming less attractive as organizations look to modernize their infrastructures.

In addition to these technical challenges, several leading data tape technology vendors such as Oracle and HP decided to halt production of their respective technologies over the past decade. This has caused ripples in the industry, with users being forced to engage in costly technology migrations ahead of plan due to these unexpected end-of-life announcements. Additionally, recent challenges emerged regarding the unstable supply of LTO media which has also accentuated long-term sustainability concerns surrounding data tape.

Due to these and many other factors, there exists a general trend away from building and maintaining large on-premise data tape infrastructures across all industries, with many accelerating moves to the cloud.

Public Cloud Services

Although not a storage technology per se, the most exciting area of advancement of digital storage technology is its availability as utility-based cloud storage services.

All major public cloud providers now offer storage services which can be effectively mapped to most or all of those highlighted previously but historically deployed on-premise and managed by the end-customers. The main benefit of these cloud-services is the treatment of *storage as a service* on the basis of SLAs rather than users having to worry about the specific technologies selected and deployed at any point in time. This means rather than focusing on the storage technology, error rates, storage density, migration timeframes, deployment locations, and other technical specifications, we need only to consider the SLA data published by the cloud-services provider. We no longer need to be concerned with what technologies they use and how they are implemented allowing us to specifically focus on business needs instead.

We recognize that each of these cloud-service providers leverages solid-state, various hard drive types as well as data tape, but you will not find this information published anywhere as they want subscribers to focus only on target SLAs rather than how they are being attained. Based on decades of experience, this is a welcome relief as we can finally focus on our target objectives rather than on the specifics of the technologies implemented, all while not ever worrying about facilities, air conditioning, power, and most importantly data migrations.

Cloud-service providers publish their SLAs in terms of nines (9s) of services such as resiliency – probability of data loss – and availability – probability of data inaccessibility. For example, an availability figure of four nines represents 99.99% time guarantee you will be able to access your data when you want it. Of course, it doesn't mean that your data will be lost but it might not be immediately available when you request it for less than one hour each year. Of course, this is the worst-case scenario provided by the SLA, so you can normally expect better results. There are much higher SLAs offered in terms of data resiliency, which is good news as this is the main parameter we care about with regard to digital archives. Cloud services typically provide an impressive 11 nines (99.999999999%) of data resiliency.

There are increasing and decreasing SLAs depending on what your objective is and how much you are willing to spend as each tier carries an associated cost. Cloud storage prices are typically metered as a per gigabyte per month fee with no cost to upload or migrate data into the cloud. This means that rather than having to plan for and fund the entire scale of your digital archive footprint at the start, as you have to do with on-premise storage technologies, you simply get started and only ever have to pay for what you consume at that moment in time. In many cases, migration to the cloud may take months or even years so being able to pay for the storage services as you consume it is a significant benefit and could even help to significantly offset associated costs when compared to building out an on-premise infrastructure.

The move to consumption-based storage is referred to as an Operational Expenses (OpEx) rather than the initial Capital Expense (CapEx) required to actually purchase and install on-premise storage hardware. This can be considered a benefit to some organizations depending on their account methods and funding models.

In addition to the two key SLA factors mentioned above, time to first byte of data or data access times vary greatly across different cloud storage tiers and must also be carefully considered. For example, the least expensive tier of Amazon Web Services (AWS) storage today is Glacier Deep Archive (GDA). This tier carries the same SLAs noted above, but it will take 12 hours to access stored data. This is certainly a long time, but accordingly, the cost is quite low and certainly something to consider when looking at long-term archive tiers as compared to on-premise data tape or tape storage shelves. What is more interesting though is that you can retrieve ALL of your data stored in GDA in the same 12-hour period whether it's a few gigabytes or many petabytes. Of course, more expensive tiers are available with much faster retrieves times, but these must be chosen carefully as consumption costs can quickly add up at scale. Currently, AWS offers more than five storage tiers each with different SLAs and retrieval times. Other major cloud providers such as Google, Microsoft, and others have comparable storage service offerings.

Planning should include detailed analysis of your real-world digital archive utilization patterns rather than be based on any assumptions as mistakes here could become costly. Data modelling and analysis is a very important first step in truly understanding your digital archive and planning your cloud implementation or transformation as it empowers data-driven decisions. The goal is to leverage these data models to accurately forecast and predict total cost of ownership (TCO) of storage technologies and/or services and ensure correct mapping against target storage SLAs before implementation. By understanding how digital archive data is used, the best combination of these many storage technologies and services can be selected to ensure the most cost-effective solution that meets SLA objectives.

In the early days of cloud, users were often concerned with the security of placing these valuable assets in a more public and less controllable environment. The truth of the matter is these cloud providers employ thousands of people focused specifically on security and resiliency, many times more than any one organization could hire themselves. The focus on security also transcends the specific needs of our industry, bringing significant benefits from other industries which are even more concerned with data security and preservation.

Cloud services offer limitless scale which also allows elastic expansion and contraction of services according to the desired objectives. Several niche cloud providers are also more directly targeting the specific needs of the media industry, tuning services and SLAs to better fit those required in digital archive environments.

A significant benefit of moving the digital archive to the cloud is having direct access to the cloud Software-as-a-Service (SaaS) ecosystem. These ecosystems include thousands of applications deployed and available on a subscription basis to build, deploy and advance cloud workflows. Having asset repositories in the cloud can enable the enhancement of asset metadata, distribution of content, asset reformatting and transcoding, launching direct-to-consumer services, and much more. These opportunities are typically too costly and complex to deploy on-premise and can now be used on a pay-as-you-go basis. It is best to fully comprehend the advantages of cloud-based services before embarking on a digital archive cloud transformation initiative, as cloud costs can easily become unwieldy if not fully understood in advance.

By embracing public cloud services, there are also more important opportunities to embrace transformational shifts in traditional content creation, collaboration, and production workflows in addition to those surrounding the digital archive itself. Users also no longer need to concern themselves with generational migrations nor do they need to worry about management, upgrades, replication, etc., as everything is inherently handled by the cloud provider.

This, however, does not change the underlying technologies utilized by these utility cloud providers who are leveraging the same technologies above but allow their subscribers to pay only for the portions they use when they use it and free up critical on-premise resources and budgets for other more core tasks.

The important thing to note here is the move to the cloud does not have to be an all-or-nothing decision. More and more organizations are slowly transforming their digital archives to the cloud, embracing a hybrid model which also spans on-premise archive technologies. These systems continue to support existing on-premise workflows while embracing cloud services and cloud ecosystem benefits where they best suit the tasks at hand at that moment.

Obviously, cost and economics typically play a large part in the decision regarding on-premise, cloud, or hybrid digital archive deployments. In most cases, the cost of deploying a digital archive in the cloud is more expensive than deploying it on-premise levering traditional storage technologies. But this does not always paint an accurate picture if you look at the long-term costs as well as the significant benefits afforded by the SaaS-based cloud ecosystem. Having your digital assets close to an infinitely scalable infrastructure with tens of thousands of apps and tools immediately available without an initial capital investment and spanning a global web of connected facilities can open up significant possibilities which cannot even be fathomed today.

Future Technologies

As mentioned previously, there have not been any truly disruptive advancements in storage technologies over the past several decades. Although significant to the industry, the launch of utility, cloud-based storage services simply leverage already familiar storage technologies.

The fortunate news is there are many research organizations around the globe actively working to disrupt the way we store data, looking at very novel approaches. Many are specifically aiming at the massive-scale, long-term, persistent, preservation grade storage market, which is a perfect fit for our digital archive future. The most interesting and compelling of these is DNA-based storage which leverages the building blocks of life to encode, store, and retrieve data:

> The global demand for data storage is currently outpacing the world's storage capabilities. DNA, the carrier of natural genetic information, offers a stable, resource- and energy-efficient and sustainable data storage solution.[20]

Several successful commercial demonstrations of DNA-based storage technology are showing promise as a viable storage technology aimed at these long-term preservation applications. They offer the promise of staggering storage densities, unprecedented shelf-life, and high data resiliency, which can allow entire digital archive collections to be maintained in a very small physical footprint without ever having to be migrated.

As storage technologies are themselves migrated toward cloud-services, it is unlikely any particular organization would choose to independently deploy DNA-based storage as part

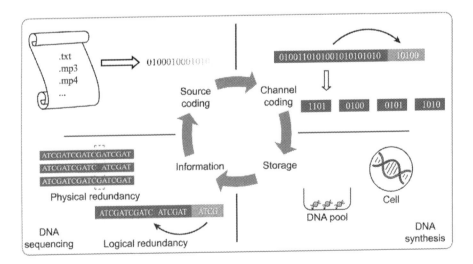

FIGURE 7.8 DNA-Based Storage (https://academic.oup.com/nsr/article/7/6/1092/5711038)

of their own digital archive. However, public cloud providers could certainly leverage this technology (or others also aiming at this preservation market) at scale to provide a very cost-effective, long-term, preservation grade cloud storage service with decades or even centuries of data resiliency and preservation.

It is obviously unreasonable to base your plans on any one public cloud provider or storage technology vendor to be in business for several centuries, so we need to consider this eventuality if our scope of protection is far reaching enough. It is therefore important we also pay attention to how our data is actually being stored, whether on-premise or in the cloud, to best assure long-term accessibility. We cover this important topic later in this chapter.

STORAGE CLASSIFICATIONS

Arguably, as entertainment and IT/enterprise technologies continue to converge, there are increasing ambiguities surrounding terminology, classifications, and technologies. In an attempt to set a baseline for the tiers of storage and their associated SLAs mentioned previously, we will take an opportunity to define each of the common storage tiers used in IT/enterprise technology and draw a correlation to those typical in entertainment and digital archive workflows.

Although the specific technologies are discussed here for context, it is important to note the corresponding cloud-services offerings are simply provided by matching target SLAs and without specific mention of the technology implemented. This further supports the paradigm shift away from the specifics of technology toward SLA-driven architectures.

Tier 0 Storage

In typical media applications tier 0 storage is usually referred to as performance or render storage and typically assigned to the most demanding and time critical operations. Entertainment workflows involving digital feature film production, ultra-high resolution editing and computer graphics render-farms often require tier 0 storage. Deployments and storage

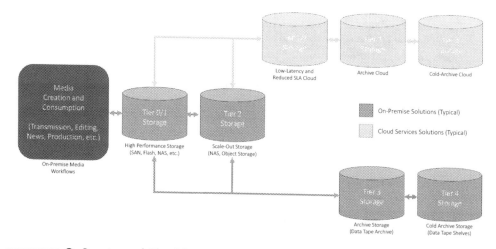

FIGURE 7.9 On-Premise and Cloud Storage Tier Definitions

capacities are often limited due to the substantial costs involved, which often push these workflows to lesser cost storage tiers. In more general, IT-centric environments:

> Business need: Extremely time sensitive, high value, volatile information needs to be captured, analyzed and presented at the highest possible speed. The primary example is currency trading. Note that this is a special-case situation not found in most business environments.
>
> Storage solution: Only storage with the highest, sub second response speeds is good enough in the currency trading environment, where a single trade can make or lose more than the cost of the entire storage system. The usual solution is solid state storage, although new high speed disk technologies may compete in this space.[21]

Tier 1 Storage

Tier 1 storage is typically referred to as *online storage* in media applications and is the most pervasive in terms of on-premise deployments. Tier 1 storage is used during for content acquisition, editing and production processes.

They are often relatively sizeable storage arrays comprised of performance NAS, SAN, and more recently flash/hybrid technologies. They often interconnect many different devices along the asset lifecycle as the usual point of play-to-air, editing and collaboration, sharing and repurposing. In IT-centric terminology, tier 1 storage is defined as:

> Business need: Transactional data requires fast, 100% accurate writes and reads either to support customers or meet the requirements of high-speed applications. One common example is online retail. Numerous surveys have shown that even relatively short delays in response to customer actions can result in lost sales, making high performance storage essential.
>
> Storage solution: Generally latest-generation, high-speed disk systems are used. These systems carry a premium price, but this cost is justified because slower performance systems would directly impact the business. However, even as disk becomes faster, solid state storage prices are decreasing, and availability is increasing. As this trend continues solid state "drives" will find their way into the Tier 1 systems of increasing numbers of organizations.[22]

This part of the lifecycle typically requires less storage than the other stages as it is normally focused on a smaller number of active productions. Because of the demands of production, these are typically performance-centric storage systems.

Tier 2 Storage

Tier 2 storage is the first that usually falls under the scope of the digital archive as it is not an active part of asset acquisition or production workflows. Tier 2 storage is typically the "scale-out" option as contrasted to significantly more expensive tier 0/1 storage. Although tier 2 storage is still relatively fast, it requires an asset *copy* or *move* operation back to tier 0/1 storage for continued production, reuse, repurposing, etc. Asset production devices do not typically directly access assets stored on tier 2 storage but rather make requests to the tier 2 storage or archive software to handle archive or restore operations to and from tier 0/1 online storage. In terms of its IT-centric definition:

> Business Need: This tier supports many major business applications from email to ERP. It must securely store the majority of active business data, where sub-second response is not a requirement, but reasonably fast response still is needed. Email systems, which generate large amounts of data, are a prime example. While users can tolerate slightly slower response times that is required for transactional systems, they are quickly frustrated by consistently slow response.
>
> Storage solution: Tier 2 technology is always a balance between cost and performance.[23]

As assets move from the creation and production phases into distribution and dissemination, they are typically moved to much larger, scalable tier 2 storage arrays sometimes referred to as *nearline storage* (NAS, Object Storage) under control of the digital archive system.

Tier 3 Storage

Tier 3 storage is most specifically in line with our digital archive objectives to ensure the long-term storage, protection, and preservation of valuable media assets. This tier has traditionally been the exclusive domain of on-premise data tape library systems, however with the evolution of equivalent cloud-services, more options exist today than ever before. In this case we will depart slightly from the IT-centric definitions as the use-cases inside our industry differ in the need for another specific, reduced-SLA disk storage tier. Therefore, tier 3 storage typically applies to data tape technologies or its cloud equivalent.

It is important to remember that in our digital archive, assets do not necessarily become less important nor do they become less utilized as they age. The CSM, Active Archive, or SDA-controlled digital archive system does not age content off tier 0, 1, or 2 storage but applies active lifecycle policies to ensure copies are maintained to meet target SLAs and ensure assets remain quickly accessible when needed. This is different in the case of HSM and backup solutions, highlighting their limitations in these digital archive implementations.

The use of near-infinitely scalable data tape archive systems or the equivalent elastic cloud-storage services allows cost-effective expansion of the digital archive system with little operational impact.

Tier 4 Storage

Tier 4 storage is by far the most cost-effective and scalable of all tiers mentioned previously and involves the storage of data tapes outside of the active tape library system on physical shelves. This is typically referred to as *offline storage* as it requires human intervention to retrieve the assets stored on these tapes. The benefit of this tier 4 storage is it gives the data tape library near infinite capacity at a very low cost, while offline assets and the data tapes that contain them are still actively tracked in the digital archive. When assets are required, a human can simply insert the tapes into the data tape library, and subsequently restored. Obviously, retrieval times vary depending on how quickly a human can find and insert the required tape, but it is typically measured in hours.

Equivalent tier 4 *cold archive* cloud-services aim to match these SLA retrieval times without the need for user intervention. Retrieval may still take hours for content stored on tier 4 cloud storage, but without the need for any human intervention or technology migrations, the advantages are clear.

Another advantage of this offline storage tier is that copies of data tapes can easily be made by the digital archive and then physically shipped to a different geographical location (i.e., Iron Mountain or other facility) for disaster recovery purposes. In the case of a catastrophic event at the main location, the digital archive can be reconstituted although not immediately as it requires the reconstruction of a physical infrastructure to insert the offline tapes into for retrieval. Although very cumbersome, many digital archive systems allow these individual offline tapes to be read to retrieve important assets to sustain operations through prolonged outage periods. This feature offers an inexpensive way to protect valuable digital archive against massive-scale loss but come at a cost of significant recovery and retrieval times.

Comparatively, cold archive cloud-services offer similar protections but require minimal intervention to bring these assets back online in the case of catastrophic failure of the primary site. Due to the cloud ecosystem and elastic nature of these services, they can also be seen to provide the additional benefit of business continuance as well. Often times, the entire digital archive can be back online and accessible in less than a day and immediately accessible to the cloud ecosystem tools to facilitate continued production, distribution, etc. As these cloud facilities are located geographically separate from the main client facilities, they offer inherent protection against local or regional events as well.

Storage Tier Summary

If we delve into some of the specific characteristics of each of these individual storage tiers, we can quickly see the advantages and disadvantages of each as they apply to specific areas of the asset lifecycle:

Storage Tier	Digital Archive Lifecycle	On Premise Technology	Cloud Technology	Relative Performance	Relative Cost	Relative Scale
Tier 0	Creation	Solid State, SAN	Cloud Workloads Only	★★★★★	$$$$$	
Tier 1	Creation	Performance NAS, SAN	Performance	★★★★	$$$$	
Tier 2	Dissemination	Scale Out NAS, Object Storage	Reduced SLA	★★★	$$$	
Tier 3	Archive	Object Storage, Data Tape	Archive	★★	$$	
Tier 4	Archive	Data Tape Shelves	Cold Archive	★	$	Near Infinite

FIGURE 7.10 Storage Tiers Comparison

You can see that cloud services offer similar tiering and distinct advantages as contrasted with each of the on-premise technologies. This is especially relevant if all workloads exist or are migrating to the cloud platform as well.

It is worth repeating that it is not an all-or-nothing decision. This is especially resonant when some existing workflows must remain on-premise while others can benefit from a cloud migration allowing users to leverage the best of both worlds. On-premise storage and cloud storage tiers can be fully integrated to seamlessly provide a hybrid asset lifecycle and digital archive solution, exploiting the distinct benefits each offers.

In the following sections, we will start to connect some of these various concepts to begin to assemble our digital archive.

THE ANATOMY OF A DIGITAL ARCHIVE

A digital archive is typically comprised of two complimentary components: software and storage. The digital archive software, or simply archive software, can be considered storage abstraction providing storage orchestration as it aims to intentionally hide the complexities of the underlying storage technologies from higher-level systems while managing automated asset movement, replication, migration, etc. The archive software typically also provides a federated, homogeneous view of all underlying storage for all connected systems and users, again abstracting the storage complexities allowing a focus on SLA target objectives. As discussed previously, there are many generations of archive software in use in the industry ranging from HSM to CSM and on to the emerging and exciting advent of modern SDA platforms.

Digital archive storage is typically comprised of different types of on-premise hardware storage technologies (solid-state storage, hard drives, and data tape archive systems) and more often cloud-based utility storage services. Again, this storage decision does not have to be all-or-nothing, as a mix of on-premise and cloud-services are typical today in hybrid storage systems. This is becoming more and more normal as the significant benefits offered by cloud-based services to next-generation digital archives are realized.

Another important characteristic to note is the digital archive software cannot dictate how the storage is being used by each of the various content production or consumption workflows but rather must support them in an agile fashion as they also morph and evolve. However, the digital archive storage does have to follow a set of very well-defined rules to ensure SLAs are met while mitigating technical factors in handling of data tapes or even moving excessive amounts of data to or from the cloud.

As highlighted, many legacy archive software systems are limited by historical factors effectively limiting the opportunity for fluid modernization to meet the fast-paced change typical in the industry today. Many of these legacy archive vendors, often in the realm of HSM and CSM, are advancing their development efforts in an attempt to support these new workflows and next-generation digital archive concepts but have ultimately moved slower than the industry placing their relevance and sustainability into question.

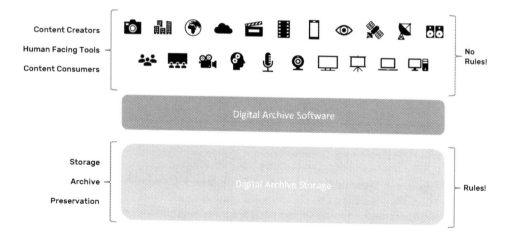

FIGURE 7.11 Anatomy of a Digital Archive

Another contributing factor to the impending demise of these legacy solutions is the parallel advancement of MAM, media supply chain, and other similar "orchestration" tools which have added the native ability to also control storage services. Once the many complexities of data tape are removed from these digital archive environments, the control of solid-state, disk, and cloud-based services becomes a much simpler task for these other systems as there are only a few standardized interfaces (S3, SMB, NFS) to most modern on-premise and cloud storage services. This makes the task of storage orchestration much easier than in the past, and on the surface, perhaps better suited to these higher-level tools, removing the specific need for digital archive software altogether. For example, MAM systems can orchestrate storage based on user-facing demands and metadata such as upcoming production, news stories and trending topics, and even enhanced and searchable metadata such as facial recognition or audio transcription information. Media supply chain systems can orchestrate storage based on production value, revenue opportunities, or direct-to-consumer demands.

However, there are two significant downsides to this approach. The first is that only the combination of the user facing tool (i.e., MAM) and the storage can be considered a digital archive, intrinsically connecting the two for the long-term. Migration, changes, or replacement of either will have an impact on the other due to this dependency. The second is the inability to effectively service other workflows not governed by these user-facing systems and the risk of having to build multiple digital archive or storage silos as a result. Ultimately, we want a single digital archive solution to traverse all parts of the media organization, servicing each natively and intimately.

Most cloud vendors offer their own simple storage orchestration solutions to manage movement of assets between their own cloud-storage tiers. These are usually quite simplistic policies similar to those found in HSM and CSM solutions based on age, last access time, or object size but could be very cost-effective and simple to manage, best suited to straightforward digital archives in the cloud. The downside of this approach is these tools do not comprehend the demands of media-centric workflows nor do they support hybrid storage or multi-cloud deployments. Essentially you have to place your digital archive and your trust in a single cloud vendor which may or may not be a point of concern.

If you are still considering data tape as a core part of your digital archive or are looking at multi-cloud approaches which are not best managed by your MAM, media supply chain, or other tools, the emerging field of SDA solutions are certainly the best option. SDA software allows assets to be stored across any number of technologies in any deployment fashion (on-prem, cloud, hybrid) and handles all complex storage interactions including replication and migration. These SDA software solutions are inherently vendor-neutral and enable intelligent and agile asset lifecycle control by metadata-driven policies which govern factors such as replication and accessibility at a SLA level.

Storage Abstraction

The storage infrastructure in the previous diagram is intentionally represented as an ambiguous block. Ultimately, we cannot foresee what storage technology fits best in the environment today nor can we attempt to contemplate the future of storage and what options may be available a decade from now. We want to endeavor to ensure our asset lifecycle workflows can continue to evolve and advance over time while maintaining a clear abstraction layer between these workflows and the underlying storage. Obviously, editors working at an edit station should not concern themselves with whether their asset is stored on data tape or in the cloud but rather rest assured their valuable content can be accessed and restored when they need it. Again, this comes down to the shift away from the intimate details of storage to the idea of SLA-driven design and architecture.

There are a few constants in the equations surrounding long-term storage and digital archives in general:

- They will employ many different types of storage technology to handle the varying demands of existing and future workflows
- They will have to migrate assets to newer storage technologies or different cloud providers over and over again during the life of the digital archive

As mentioned previously in this chapter, these are not new concepts nor requirements and have little to do with the specific technologies selected today or tomorrow. Storage technology choices will likely change more often than the workflows and user-facing tools, devices, and interfaces. It is a core tenant of modern digital archives to handle addition, modifications, removal, and migration of underlying storage technologies transparently with little or no risk to higher-level workflows, and certainly no risk to the digital assets themselves. The digital archive must abstract the complexities of the underlying storage technologies and provide a federated and vendor-neutral view regardless of what storage technology or cloud-service is utilized.

Storage Orchestration

Storage orchestration relates to intelligent and automated asset movement, migration, and replication under the hood of the digital archive software without specific guidance or commands from higher-level systems. Storage orchestration essentially relates to the SLA targets we define and govern how and when content is replicated, moved or migrated, and to which specific target storage devices. For example, an orchestration policy may be defined to ensure a copy of a specific asset class or group is maintained in tier 2 on-premise storage while a parallel copy is made to the least-expensive public cloud tier for disaster recovery. This cloud instance may reside in tier 2 or tier 3 storage for fast access while the production is underway. Once production is complete, it may be automatically deleted off on-premise tier 2 storage and moved to tier 4 cloud storage to minimize storage costs. This same asset could then be intelligently moved back to tier 2 cloud storage based on metadata which defines the production schedule to ensure it is immediately accessible when needed, intelligently overcoming the access latencies for tier 4 cloud storage.

SDA solutions can handle these complex storage orchestration tasks and fundamentally highlight the shortcomings of legacy CSM systems. To empower the paradigm shift towards SLA-driven digital archives, it is imperative these orchestration decisions be based on metadata rather than file or object characteristics typical in these legacy systems.

Archive Maintenance

Historically, film and video tape were used as a storage medium for movie productions, news stories, and episodic content. Most content creators amassed many thousands of these physical assets over many decades of production on several generations of video tape and film technologies. Over time, these legacy video tape and film formats required periodic migration to newer technologies for fear the physical player devices would no longer be available to retrieve the assets. This historical burden was immense as every hour of video tape and film would take at least one hour or more to migrate to a new format. Additionally, this *dubbing* process typically involved a human to monitor the content during the entire process to ensure the best-possible outcomes, as the original medium would be subsequently disposed of or become inaccessible due to the player machines becoming unavailable. Each time this asset was dubbed, it also went through a *generational loss* effectively reducing its quality by a small amount each and every time.

When dealing with dozens of hours of content, this migration process was not terribly difficult, costly or time consuming. However, when you scaled an asset collection to thousands or even hundreds of thousands of hours, these migration tasks became almost perpetual and of the scale where few organizations could afford the time or expense. As a result, there are still thousands of important video tape and film collections around the globe at risk of loss due to this insurmountable migration time and expense challenge.

Some innovative content owners recognized the value of these legacy physical archives and embarked on large-scale asset digitization processes. This represented an investment to move asset collections away from their legacy physical storage media and into the realm of assets managed by digital archives.

Periodic asset migration inside the digital archive is typically fully automated as mentioned previously. It often involves no human intervention aside from simply initiating the migration process. As these assets are moving as "bits" of data in the digital archive, automatic integrity checks can be performed on the data to ensure the newly migrated instances are exactly the same as the original one eliminating the costly human quality assurance step. We also benefit from the fact that these are data migrations rather than audio/video migrations, meaning there is no generational loss each time a copy is made as the copy is an exact and authenticated replica of the original instance. As the performance characteristics of modern storage technologies far surpass the "realtime" bounds we encounter when dealing with legacy video tape and film formats, these migrations can occur many times faster than realtime overcoming the challenges of data scale. Lastly, the periodic migration and processing of these digital assets offers a window of opportunity to perform fixes, enhance metadata, validate integrity, increase storage densities and performance, and even embrace new storage technologies and cloud services.

Asset Integrity

When digital assets reside on spinning disk of any sort whether on-premise or in the cloud, it is quite easy for modern digital archive software to automatically validate their integrity. This can be achieved by simply running a checksum calculation when system resources are available or whenever an asset is being restored, moved, or replicated.

This asset integrity-check process assumes a "known good" value for this checksum is stored as metadata along with the asset upon its inception. Asset instances where corruption is discovered can simply be automatically removed and replaced with an instance which has been properly validated residing on another storage system. In cases where a valid instance is not available, human intervention may be needed to leverage digital triage tools to salvage whatever data is still accessible, so it may not represent a total loss. Digital archive best-practices dictate a minimum of two copies of all assets are maintained at all times in geographically separated infrastructures. It is important to note that by making additional copies or instances of assets in the system, their resiliency SLA values combine better ensuring their long-term protection against loss.

Asset integrity checks become much more difficult if your assets are stored on data tape in a tape library system as they are not online and readily accessible at all times. There are competing philosophical camps on whether it is better to periodically load data tape media into tape drives to perform integrity checks on the data or leave them safely sitting inside the tape library peacefully untouched. It is true that loading a data tape is quite a mechanically intensive operation with a fast-moving robotic arm grabbing the tape, transporting it, and loading it into a receptive tape drive. Once inside the drive, the tape needs to be mechanically threaded and wound around a reel inside the tape drive itself, quickly shuttled to the location of the requested data, tape speed and tension stabilized, and the read process started. Despite careful engineering and awe-inspiring design, these mechanically heavy operations could be catastrophic to the sensitive data contained on each tape cartridge. Leaving them peacefully sitting inside the tape robotic system and sparing them the mechanical interruptions may better protect the data contained.

Generally, it is not recommended to sporadically load data tapes into a drive for the sole purpose of validation and integrity checks unless of course the media itself is suspect. The recommendation is to simply leverage the randomized nature of asset access typical in these digital archive environments to statistically verify assets which are contained on cartridges loaded to service other read or write operations. If we consider setting a threshold age of one year (or other somewhat arbitrary figure), there will be "qualifying" assets contained on tapes which are loaded into drives to service actual production requests. By selecting one or more of these qualifying assets contained on tapes which are already loaded, we minimize unnecessary mechanical wear and tear on the tape media and still accomplish our goal of spot-checking validity of assets over time. Spot checking aged assets on media will give a good indication of media and assets at risk and help prevent large-scale asset loss.

Storage migration to newer technologies every decade also enables full asset validation to occur in-path while opening up the additional opportunity for triaging, replacing, and rebuilding failed instances or even making cloud copies.

Connecting the Bits and Pieces

Being able to select and manage any storage type and technology is a wonderful benefit provided by the digital archive. However, as important is the tight integration into the production tools, services, devices, and interfaces used as part of the media asset lifecycle. Our ultimate objective is to have the digital archive fit the creative process and business needs rather than these conforming to fit the digital archive.

Modern SDA solutions expose heterogeneous and globally federated views of the underlying storage via industry standard protocols such as S3, SMB, and NFS. This allows both simple and complex connections to high-level systems such as MAM, media supply chain, and others without them having direct awareness of the underlying storage technologies employed. This allows the digital archive to remain separate and independent from the tools which use it, allowing these tools to be added, removed, upgraded, or decommissioned without affecting the integrity or operations of the digital archive.

Modern SDA platforms are the *glue* between various human-facing tools, devices and interfaces, and the complex and ever-changing digital archive storage infrastructures.

ARCHIVE TRANSFORMATION AND MODERNIZATION

With all of these many complex factors taken into consideration, the truth of the matter is that many organizations already have digital archives in place and are faced with specific challenges surrounding their modernization. Depending on whether these digital archives were implemented in an ad-hoc, siloed fashion or as enterprise-class digital archives, there is likely desire or impending need to take a step toward a next-generation digital archive. The ultimate objective is to always focus investment on the future of an organization rather than on its past.

From a digital archive software perspective, the legacy vendor may still exist and continued development of their system to adapt to some of the modern demands of the industry. Careful assessment is required to determine whether the best forward-looking plan is to simply remain status-quo and upgrade existing software to ease the implementation of newer technologies and services or take a more aggressive step to abandon it in favor of more modern solutions or services. If the legacy archive system supports scaling to the cloud, this may be the easiest choice to make as long as vendor lock-in and proprietary interfaces can be avoided, which is often difficult to ensure. Significant investment in time and money could land an organization with a modernized but no more sustainable digital archive solution.

From a storage perspective, these large-scale legacy archive systems almost exclusively rely on aging data tape infrastructures as this has been the defacto standard for close to two decades now. Driven by inertia, the path of least resistance may again be to invest the capital to secure data center space, purchase a new data tape library, data tape drives, and data tape media and perform data migration under legacy archive system control to increase performance, storage density, and asset longevity. Again, this money is likely better spent on a more aggressive cloud transformation plan rather than simply deferring the inevitable migration away from data tape and its complexities.

However, the biggest archive modernization mistake organizations can make is to move from one digital archive software vendor platform managing on-premise data tape to another digital archive software vendor platform also managing on-premise data tape. It is much less risky and less expensive in the long-term to stay with the legacy solution proven in the environment than risk an archive transformation only to end up in much the same place at the end of the migration. Reducing continued investment in these aging legacy archive platforms while planning a more aggressive archive transformation is the best path to the next-generation digital archive. It is much better to simply wait until strategies and budgets support a true archive transformation than to continue to make substantive investments in these legacy systems.

Working toward this end goal, organizations can almost immediately begin to alleviate reliance on legacy archive applications and on-premise data tape infrastructures. In some cases, this could involve a modest upgrade of the legacy archive software to support orchestrating over existing on-premise data tape as well as cloud services. This can help to cloud-enable the digital archive without the necessity to refresh data tape storage infrastructures. This may allow an organization to continue forward with a diminishing reliance on on-premise data tape, begin to realize the benefits of cloud-services, and slowly move away from on-premise storage as it ages out of the environment. Even if this slow migration takes several years to complete, money can be redirected from capital intensive on-premise storage systems to fund increasing cloud adoption. Once the cloud migration is complete, organizations can then determine whether the legacy archive system is still relevant in this new environment, whether it can be replaced or just turned off altogether. Caution is advised as legacy archive vendors are specifically motivated to ensure they maintain control over existing deployments and the digital assets they store, so careful and thorough advanced planning is required to avoid long-term lock-in.

Thankfully, migration solutions exist today which avoid costly and risky upgrades of both legacy archive systems and their underlying storage infrastructures while ensuring no long-term vendor lock-in. These advanced archive migration orchestration solutions can perform fully automated, background asset migrations to the cloud while maintaining uninterrupted legacy archive operations on-premise avoiding costly upgrades altogether. These archive migration solutions effectively *borrow* some portion of the existing legacy archive system and storage resources to facilitate background archive migration. They also federate and abstract the legacy archive and cloud storage services, effectively making both appear as one *virtualized legacy archive* to all existing applications already integrated with the legacy archive (i.e. MAM, automation, traffic, etc.). This helps avoid costly and risky upgrades to these business systems and human facing tools, effectively decoupling them from the digital archive now and into the future. During migration, assets are also freed from any proprietary containers or wrappers ensuring all assets arrive in the cloud in an open and accessible format, ready to operate fully independently of the legacy archive system. During and after the migration, migration orchestration solutions ensure on-premise legacy archive systems and cloud storage services remain fully synchronized until the legacy archive system is finally decommissioned. As the legacy archive system continues to operate in parallel during the migration, and for some period afterward, it can be considered a backup or *cloud safety-net*, allowing a risk-free digital archive transition to the cloud.

Obviously, there is no one-size-fits-all approach to archive transformation. It is recommended that careful planning and consideration be given to the objectives of the organization. It may also be helpful to engage with various subject matter experts (SME) in the field of digital archiving and to leverage the experience of others facing similar journeys.

At a simplistic level, some questions to ponder when planning a next-generation archive transformation:

- Are we ready to move off data tape?
- Are we facing a fork-lift upgrade of our legacy archive system or storage in the coming years?
- What continued value does our legacy archive system bring to the organization?
- Can we consider cloud services on a scale larger than just that of our digital archive storage?

In many cases, the largest barrier to archive transformation is the inertia of the status-quo as these legacy archive systems are often tried and true, serving the needs of the organization successfully for many years. It is important to remain focused on the key strategic objectives of the organization and ensure alignment with the plans around digital archive system modernization and transformation. It is also important to remember that these massive-scale migrations can take months or even years to complete, so any planning must take this into account.

Regardless of whether an archive transformation is in the near-term plan, it is recommended that some effort be placed into fully understanding and quantifying sustaining legacy archive costs and risks each year. By regularly assessing factors such as the age of the infrastructure, impending capital intensive upgrades, migration timeframe estimates, sustaining annual hardware and software support costs, tracking legacy archive vendor viability and the expense of skilled employees dedicated to maintaining these infrastructures, organizations can be better prepared to ensure a pragmatic and educated path to their next-generation digital archive.

STANDARDS-BASED DIGITAL ASSET PROTECTION

Despite all of the considerations necessary in planning and architecting a digital archive system, it is important we also look at the way these valuable assets are stored on the storage systems to best assure long-term accessibility and ensure independence from the archive software itself. This is often referred to as vendor-neutrality and is an important objective in ensuring long-term asset access. There are obvious benefits to storing assets in their native format, but arguably this can limit advantages of other approaches as important metadata and context can be missing.

Important standards development work has been underway over the past decade to embrace the modernization of technologies and approaches to this ongoing preservation challenge. These standards attempt to combine the desire for native access while layering on important metadata-encapsulation and preservation features to better assure long-term viability and access to digital assets. The good news is experts in the field of storage, archive, and preservation have worked with important global standards bodies such as the Society of Motion Picture and Television Engineers (SMPTE) as well as the International Standards Organization (ISO) to solidify this important topic as globally published, international standards.

Archive eXchange Format (AXF) Standard

AXF is an open-standard object-container for digital data and its associated metadata. It allows any type, size, and number of files or data blobs to be stored, transported, and preserved as a collection on any type of storage technology while maintaining full independence of file or operating system and underlying storage. Over a decade in development, AXF addresses the need for a long-term open storage and preservation format while overcoming the limitations of legacy file and container offerings. AXF was first published as an international standard by the SMPTE in 2014 and later by ISO/IEC in 2017.

AXF bundles any type and amount of essence data and associated metadata (structured, unstructured, descriptive, technical, proprietary, etc.) into an immutable, self-describing preservation package while maintaining the ability to dynamically interact with any of the data elements contained inside an object. Its objective is to ensure valuable assets will be protected now and into the future, regardless of the technology you choose today or may choose tomorrow in your digital archive.

AXF is the physical embodiment of the object store concept targeting the transport, storage, and long-term preservation of complex digital asset collections. It is a revolutionary way for digital archive systems to package essence and metadata, extending the encapsulation to the storage devices and technologies they maintain.

AXF is like an advanced ZIP[24] package which encapsulates any type and number of data elements (files, folders, data blobs, etc.) in a very well-structured, self-describing data object. An AXF Object carries any type and amount of metadata along with the data payload. It also includes redundant index structures and an embedded file system. This embedded file system ensures independence from any filesystem, operating system, or storage technology on which it is created and stored. This means AXF Objects are universal and can be stored on any type of technology including data tape, spinning disk, flash media, and even the cloud. Because the AXF Object structure is very tightly defined, it is always the same regardless of the technology employed removing the complexity and risk associated with the selection of storage technologies. AXF also adds self-describing characteristics to the media itself, which ensures that it can be accessed and read regardless of whether the digital archive system which originally created it and the AXF Objects it contains is available or not.

AXF is a streaming-centric technology, also being used extensively for the transport of complex file collections across WAN networks as well as to and from the cloud. Because each AXF Object fundamentally connects metadata and the data payload which comprise the asset in an encapsulated container, important context is always maintained. It also includes preservation-centric features and allows the receiving system to track the chain of custody (or provenance) of the asset, validate structures and integrity of data and metadata on-the-fly, and ensure authentication and maintain access control for the asset collections being transported and stored.

FIGURE 7.12 Archive eXchange Format Universality

Key to the adoption of AXF is its nature as an application-level implementation which does not require the storage vendors do any development to support it. In addition to this, AXF does not rely on any specific or proprietary features of the storage technology used, so it is supported in current as well as older generation technologies. It is currently being used in large-scale digital archive environments across diverse storage and network transport technologies supported by multiple technology vendors.

AXF was specifically developed to address the need for a long-term open storage and preservation format and overcome the limitations in legacy container formats such as Tape ARchive (TAR[25]). This format was often used by archive systems in the past in order to achieve some level of vendor neutrality and interoperability but lacked key features required for digital archive storage and preservation. AXF technology abstracts the underlying file and operating system technology facilitating long-term portability and accessibility to the files contained in the container while adding several Open Archive Information Systems (OAIS[26]) preservation characteristics to provide long-term protection for digital assets.

AXF defines the specific method for encapsulation of file or data collections and their associated metadata as well as the mechanism for storage on media storage technology (data tape, magnetic disk, optical media, flash media, cloud, etc.). AXF adds self-describing features to both the AXF Objects and the media on which they are stored, allowing independence from the systems which originally created them. AXF provides a universal and open file-system view of all stored objects, data, files, and metadata allowing exchange with any applications which also comply to the published and maintained international standards.

AXF takes the concept of an object store to the physical level as a fully self-describing, self-contained encapsulation format for complex data collections. AXF provides a standardized way of storing data, files or file collections of any type and size, along with limitless encapsulated metadata of any size and type, on any type of storage technology or device. AXF is independent of the host operating system and file system as well as from the application that originally created the AXF Object ensuring vendor and technology neutrality.

A Closer Look

We have covered much about the architecture and technologies involved in digital archive operations inside the media and entertainment industry. Arguably, more important than where or what is used to store digital assets are the way they are stored on the selected storage technologies or services.

AXF is based on a file and storage medium-agnostic architecture which abstracts the underlying file system, operating system, and storage technology. AXF Objects contain any type, any number, and any size of files as part of its payload, accompanied by any amount and type of structured or unstructured metadata, checksum and provenance information, full indexing structures, and other data within a single self-describing, encapsulated package. The AXF Object includes an embedded file system, which helps abstract complexities and

FIGURE 7.13 The AXF Construct

FIGURE 7.14 AXF Objects on Storage Media

limitations of underlying storage technology, file systems, and operating systems. AXF Object can exist on any data tape, disk, flash, optical media, cloud or other storage technology and can be used for network transport of data.

As a self-contained and self-describing container, AXF supports large-scale digital archive systems as well as simple standalone applications, facilitating encapsulation or wrapping, long-term protection, and content transport between systems from various vendors which conform to the AXF standard.

AXF is an IT-centric implementation, supporting any type of file or data encapsulation including databases, binary executables, documents, image files, etc. It supports the OAIS model as well as preservationist features such as provenance for both media and objects, unique identifiers (UUID, UMID, ISCI, etc.) support, geo-location tagging, error detection down to the file, and structure level and data-validity spot checking.

AXF Structures

Embedded File System

The AXF standard leverages its own embedded file system. AXF offers a translation between any type of generic file or data set and logical block positions on any type of storage medium being used with or without its own file system. AXF encapsulates a related set of files or data elements with any type and amount of ancillary metadata (structured or unstructured) into a single, well-defined container.

AXF is intended to overcome the limitations of other encapsulation formats, which cannot support complex file structures with millions (or more) of files or handle large assets sizes typical in media and entertainment particularly well. Because of its embedded file system, AXF does not depend on the storage technology. Although optimized for the storage of large assets and collections, the AXF format can be applied to any environment where an open and accessible storage encapsulation format is required for any type of asset collections.

Object Encapsulation

Each AXF Object is a fully self-contained, encapsulated collection of files, data, metadata, and any other ancillary information which adds relevancy or value to its contents. AXF is designed to handle a single file encapsulation or many millions of elements of any type and size.

AXF Objects are equivalent regardless of whether they are created on data tape, disk, flash, cloud or future storage technology with or without file systems. This makes the creation and handling of AXF Objects on differing media deterministic, better ensuring long-term accessibility.

Structures

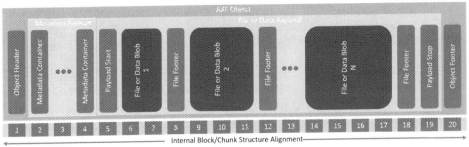

FIGURE 7.15 AXF Object Internal Structures

Block and Chunk Alignment

To aid in resiliency and performance across any storage device, technology, or medium, each data structure and element contained within an AXF Object is aligned on pre-defined chunk boundaries. These can be independent of the storage technology or medium and can be different for each AXF Object contained on that medium if the medium itself supports this. Chunk size is typically an integer multiple or sub-multiple of blocks on block-based media which also aids in the recoverability of AXF Objects where indices may have been damaged or corrupted.

During each AXF Object creation, copy or movement operation, the application is responsible for ensuring the data is aligned on the block boundary basis defined for the destination storage technology or media. AXF Objects can be specifically tuned to the underlying storage to optimize performance and storage efficiency even down to the level of tuning on an AXF-object-by AXF-object basis, depending on the average, minimum, and maximum sizes of data elements contained within the data payload.

VOL1 Identifier

This VOL1 Identifier structure is included only on data tape media when used to store AXF objects. It is included to identify the tape as AXF Media to other applications and systems which might also try to access these data tapes to ensure they are properly identified. It is not relevant to any other AXF implementations or storage technologies.

Medium Identifier

The AXF Medium Identifier structure contains the AXF volume signature, a UUID, and label for the media as well as information about the storage medium itself. The implementation of the Medium Identifier differs slightly depending on whether the storage medium is linear or nonlinear, and whether it includes a file system or not, but the overall structures are fully compatible.

This structure adds the self-describing media characteristics to AXF so systems that support the format can immediately understand everything necessary about the media to be able to index, read, and recover its contained AXF Objects.

Object Index

The AXF Object Index is an optional structure that assists in the rapid recoverability of AXF-formatted media by foreign systems. Information contained in this structure is sufficient to recover and rapidly reconstruct the entire catalog of AXF Objects on the storage medium – think of it as an advanced File Allocation Table (FAT[27]).

This Object Index can be periodically written at various points on the storage media or updated periodically on file-based systems to provide enhanced, rapid indexing and recoverability. In a case where the application has not maintained these optional AXF Object Index structures, the contents of each AXF Object can still be reconstructed by rapidly processing each AXF Object Footer structure adding to the multilevel enhanced resiliency of the format.

Inside an AXF Object

Binary Structure Container

The Binary Structure Container is a simple binary envelop which wraps/contains payload information and includes structure identification, index structure checksums, classification information, media/mime types, etc. For AXF structural elements, this payload is simple XML, but in the case of Generic Metadata Containers, it can be anything from binary to text or even source code. The Binary Structure Container allows the application to comprehend its contents allowing it to be stored, validated, tracked, and reliably recovered regardless of its nature or origin. An important characteristic of AXF is that the files or data elements contained in the actual data payload of the AXF Object do not require Binary Structure Containers and are simply bit-for-bit equivalents of the original data, not modified in any way. This is an important feature for preservation in that the asset data is never modified in any way and ensures that even the simplest AXF applications can read and retrieve the data payload from an AXF Object.

Object Header

Each AXF Object begins with an AXF Object Header structure. This structure contains descriptive XML metadata describing the actual contents of the AXF Object such as its unique identifier (UMID/GUID), creation date, descriptive information, file tree information, permissions, etc. This acts as a detailed *table of contents* or catalog for the data contained

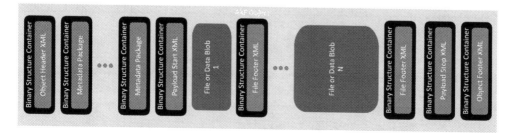

FIGURE 7.16 AXF Universal Binary Structure Containers

inside the AXF Object. By simply parsing this structure, AXF applications can comprehend everything about the asset data, where it came from, what was used to create it, how to retrieve it, etc. This structure contains several mandatory metadata and descriptive elements to ensure long-term accessibility and interoperability between AXF systems.

Metadata Container

Following the AXF Object header are any number of optional AXF Generic Metadata packages. These are self-contained, open metadata containers for applications to include AXF Object-specific metadata. This metadata can be structured or unstructured, open or vendor specific, binary or XML, and provides a flexible and dynamic space in which to enrich the depth of the AXF Object and permanently link it to the encapsulated data payload of the object.

There are no constraints or standards governing the type of metadata, the number of packages, or their contents. In the case where there is no metadata to store along with the AXF Object, this structure is simply omitted. Metadata (XML, binary, or other) can be stored in the file payload of the AXF Object as well, but their context may make processing difficult for third-party applications during subsequent restore or replication operations and are often better suited to be carried in the metadata payload section of the AXF Object.

File or Data Payload

Next is a Payload Start structure which marks the start of the essence data payload of the AXF Object. This is a simple, empty structure which can be easily located by even the simplest AXF applications.

The file or data payload of an AXF Object consists of zero or more Data + Data Padding + Data Footer triplets. This is the actual byte data of the files to be stored within the AXF Object container. File Padding is used to ensure the chunk alignment of all AXF Object files and structure elements which is a fundamental requirement of the AXF standard. The File Footer structure contains the exact size of the preceding file along with file-level checksums, original path information, etc. and can be rapidly processed by the application on-the-fly during operations to retrieve files, validate authenticity and ensure data integrity.

Following the last data elements in the payload, a Payload Stop structure marks the end of the data payload of the AXF Object. This is a simple, empty structure which can be easily located by even the simplest AXF applications.

Object Footer

The final portion of an AXF Object is the AXF Object Footer. It is essentially a repeat of the information contained in the Object Header with some additional information captured during the creation of the AXF Object itself such as file checksums, block positions, and file permissions. The Object Footer is fundamental to the resiliency of AXF and allows efficient re-indexing of AXF media by foreign systems when the content of the media is not previously known (i.e., the self-describing nature of AXF Objects).

Following the initial creation of an AXF Objects, any subsequent move or copy operations mandate the information contained in the AXF Object Footer be fully replicated in the

AXF Object Header (if this is not already the case), adding to the redundancy of the AXF standard.

History of AXF

The Early Days

In July 2006, the SMPTE V16-ARC AHG initiative was formed by the three leading digital archive vendors at the time (Front Porch Digital, Masstech and SGL) to focus on the invention and development of an open-standard for the storage and preservation of digital assets on any type of storage device or technology. Merrill Weiss was selected as the SMPTE committee chair to lead this initiative.

In October 2008, the V16-ARC AHG received status as an official working committee under SMPTE and was renamed the 10E30WG-AXF or the AXF working committee. By April 2008, Front Porch Digital was the only remaining founding member in active committee participation. In October 2009, the SMPTE AXF committee ceased activities due to limited participation and the AXF initiative was shelved.

In appreciation of the importance of AXF and the objectives it embodied, Brian Campanotti (CTO, Front Porch Digital at the time) brought the initiative inside of Front Porch Digital to continue its advancement. He formed a small working group including users and members of his development team Grégory Pierre and Benoît Goupil. The original technical framework was abandoned and AXF was reinvented from the ground up while maintaining the key industry and end-user objectives as defined by the original SMPTE committee and by the industry.

Release to the Market

At the National Association of Broadcasters (NAB) conference in Las Vegas in April 2011, Front Porch Digital announced the completion of their AXF implementation and its release to market as part of their DIVArchive V7.0 Content Storage Management (CSM) solution. Brian Campanotti further announced their AXF specification and design would be contributed to SMPTE as an unencumbered straw-man proposal for continued advancement and eventual standardization by the AXF committee. At the same time, an AXF community website (www.axf.io) was launched for vendors and end-users interested in AXF technology to gather industry input and feedback on the invention.

The SMPTE 10E30WG-AXF committee (previously under the TC-10E Essence parent technical committee) was placed under the TC-31FS File Formats and Systems parent committee. As a result of this reorganization, the official name of the SMPTE AXF committee was changed to TC-31FS30 AXF WG, and this straw-man specification formed the basis for its revitalized work.

AXF Becomes a Standard

In September 2014, it was announced that AXF was ratified and published as SMPTE standard ST 2034-1:2014 and is available for direct download from the IEEE Digital Library.[28]

As part of a personal commitment made in 2011, Brian Campanotti spearheaded an effort to bring the new SMPTE AXF standard to ISO/IEC for standardization to help broaden the reach of this important technology. The ISO/IEC PAS submission process was initiated in March 2016.

The SMPTE AXF standard was submitted to ISO/IEC a year later, and following a two-month translation phase, AXF entered balloting on June 7, 2017. After a two-month balloting period, AXF was ratified in August 2017 as ISO/IEC 12034-1 Standard with no objections from any voting nation. This was a very important moment for the entire AXF team and the industry, eleven years after the original conception of AXF. The ISO/IEC 12034-1 AXF Standard is available for download directly from the ISO/IEC website.[29]

AXF Adoption

AXF has been successfully deployed across various storage technologies (flash, disk, tape, optical, and cloud) and a multitude of file-based asset types in many industries touching various applications and environments. In addition to storage and preservation, AXF is used by several global organizations facilitating authenticated and protected WAN-based transport of valuable assets between globally distributed facilities, as well as to-and-from multiple cloud service providers.

AXF has been successfully deployed on all major storage technologies in industries spanning from national archives, museums, government agencies, film production, and broadcasting to generic IT compliance archiving. It is estimated there are hundreds of petabytes of the world's most valuable (monetarily, historical, cultural) digital assets stored, protected, and preserved by AXF – a true testament to the dedication and hard work of many companies and individuals over the past decade.

AXF Today

The SMPTE AXF Committee and the axf.io community remain active today. Tracking and assisting development organizations and end-users with their AXF implementations, collecting requirements and suggestions for expansion and improvement. In addition to this, work is underway on AXF Part 2 which extends the reach of AXF earlier in the digital asset creation lifecycle. Essentially, AXF Part 2 is a "de-encapsulated" version of AXF Part 1, allowing for each component to be created, manipulated, edited, versioned, etc. during its lifecycle, while capturing important metadata and provenance information essentially reaching back to the birth of the asset. Obviously, a large part of AXF Part 2 is the ability to map bidirectionally between AXF Part 1 and Part 2.

This work extends the relevance of AXF from the birth of a digital asset through to its eternally preserved life, capturing protecting and preserving as much as possible along the journey.

CONCLUSION

There are many complex factors to be considered when looking to implement or modernize an existing digital archive system. The shift from purpose-built storage silos deployed and evolved over time to a modern, federated digital archive can represent a monumental shift for an organization having to consider requirements from stakeholders across the organization. The challenge is further exacerbated for organizations which span many facilities or even many countries and likely possess very different views on the digital archive.

Careful analysis backed by real world data is the best first step toward fully comprehending the requirements and associated costs and can help to point organizations in the right direction when planning this important endeavor. Although we often head into these projects with lofty goals and objectives, costs will ultimately be the primary driver of decisions and even many concessions along the way.

Modern cloud-services have enabled us to look at digital archives from a service level agreements (SLA) perspective rather than bother with the specific pros and cons of each specific storage technologies and how they fit into our architecture and our budgets. This fact empowers organizations to take a business driven, asset-centric approach to their digital archives rather than the traditional and much more complex storage-centric approach.

However, for many organizations the shift from CapEx to OpEx will present financial challenges which often times are not fully understood at the onset. Careful TCO planning, modeling and analysis combined with a solid understanding of the objectives can empower accurate and predictable cloud cost control.

Organizations must understand the importance of development, funding and sustaining a solid digital archive strategy and its importance to future value, growth and agility. As viewer and consumption demands change, it is an important enabling technology that empowers organizations to be ready to react and pivot to ensure growth, relevance, and a viable business future.

Today, cloud-services enable a level of agility for organizations never before seen. New direct-to-consumer services can be quickly launched leveraging the assets of an organization in a very cost-effective fashion. By rapidly assessing viewer reach and return on investment, these new services can be quickly adapted, tuned, expanded or even cancelled. Because of cloud pay-as-you-go pricing, financial exposure in these cases is minimal when compared to the capital and time intensive launch of new on-premise services in the past. Key to this agility is having the digital assets readily accessible and close to the tools and services which will leverage them in the cloud, which is simply not possible or practical if on-premise data tape infrastructures continue to warehouse digital assets. Although proven over the past two decades as the most reliable and cost-effective digital archive storage, these aging approaches to digital archives are falling by the wayside in favor of more modern architectures and cloud-centric services due to the significant benefits they provide.

The choice of digital archive software is also a daunting challenge but can be easily assessed once the goals and objectives have been defined. Obviously, there is no need to introduce additional complexities and vendor systems if there is not a clear advantage to the organization. In some cases, organizations will choose to solely rely on tools and services provided by their selected cloud vendor. For some, they may choose to proceed with MAM or media supply chain systems to orchestrate assets across their storage tiers. For others, a modern digital archive approach enabling cognitive metadata-driven asset orchestration leveraging SDA technologies may be the best fit. Regardless of the forward-facing plan, it is clear that money invested in aging legacy archive systems and its related storage infrastructures is not money well spent.

No one would generate, acquire or produce content unless there was some perceived long-term value. It is important that decisions be made carefully and backed by data and analysis to ensure the best chance of long-term availability and accessibility to valuable digital assets.

Lessons should be learned from those who have gone before, and no one should assume their objectives differ materially from others grappling with similar challenges and hoping to achieve the same digital archive objectives.

As technology continues to advance, best practices should be leveraged to ensure the most risk-free path is chosen. Leveraging modern storage and cloud-services natively or enabled by advanced SDA digital archive solutions can best assure a trouble-free path to long-term storage, archive, and preservation now and into the future.

Digital archives store and protect the crown jewels of any organization and must be attended to with care.

NOTES

1. https://en.wikipedia.org/wiki/Library
2. https://en.oxforddictionaries.com/definition/repository
3. https://en.oxforddictionaries.com/definition/archive]
4. https://www.merriam-webster.com/dictionary/preservation
5. https://en.wikipedia.org/wiki/Archival_science
6. https://www.popularmechanics.com/technology/security/a28691128/deepfake-technology/
7. https://en.wikipedia.org/wiki/Digital_preservation
8. https://en.wikipedia.org/wiki/Open_Archival_Information_System
9. https://www.oxfordlearnersdictionaries.com/definition/american_english/asset
10. https://news.un.org/en/story/2019/10/1050081
11. http://www.unesco.org/new/en/communication-and-information/access-to-knowledge/linguistic-diversity-on-internet/initiative-bbel/stakeholders/unesco-strategy-for-the-safeguarding-of-endangered-languages/
12. https://sfi.usc.edu/what-we-do/collections
13. https://aws.amazon.com/solutions/implementations/media2cloud/
14. https://www.idc.com/getdoc.jsp?containerId=prUS46286020
15. https://en.wikipedia.org/wiki/Stub_file
16. https://en.wikipedia.org/wiki/Content_storage_management
17. https://en.wikipedia.org/wiki/Active_Archive_Alliance
18. https://cloudfirst.io/sda.html
19. https://en.wikipedia.org/wiki/Service-level_agreement
20. https://academic.oup.com/nsr/article/7/6/1092/5711038
21. http://wikibon.org/wiki/v/Five_Tier_Storage_Model
22. http://wikibon.org/wiki/v/Five_Tier_Storage_Model
23. http://wikibon.org/wiki/v/Five_Tier_Storage_Model
24. https://en.wikipedia.org/wiki/ZIP_(file_format)
25. https://en.wikipedia.org/wiki/Tar_(computing)
26. https://en.wikipedia.org/wiki/Open_Archival_Information_System
27. https://en.wikipedia.org/wiki/File_Allocation_Table
28. https://ieeexplore.ieee.org/document/7879152
29. https://www.iso.org/standard/73525.html

CHAPTER 8

Looking Ahead

Section Editors: Chris Lennon & Clyde Smith

We conclude with a section discussing where things are likely headed in the area of media workflow. This is an area that is undergoing constant change, and so understanding where things are going is the most important thing, once the reader has a good grasp of the entire ecosystem. Industry trends, as well as future specifications and standards that will help to make the media workflow puzzle a much easier one to bring together will be addressed.

We have enlisted industry experts Al Kovalick and Stan Moote to peer into their crystal balls and tell us what they see.

LOOKING DEEP INTO THE FUTURE: AL KOVALICK

With 100+ years of SMPTE's technology and standards developments behind us it is fitting to look ahead and ask, "what could the future look like for video technology and facility infrastructure?" On July 1, 1941 the first commercial TV stations WNBT and WCBW made use of the new NTSC video and over-the-air standards to start broadcasting in New York City. The video signal was analog, 525 line, interlace, black & white and at exactly 60 (not 59.94…) fields per second. In 2020 we have all digital, UHD/4/8K progressive images with high dynamic range, wide color gamut, and high frame rate choices. The image quality is several magnitudes improved from the early days. Consider too the primitive systems for analog video production in 1941 compared to the all-digital, IT-based systems of today.

Over this 80-year period, many notable improvements would have been impossible to predict. So, is it a "fool's errand" to predict what video technology and infrastructure design may look like in 2036, 15 years hence? Observe what the famous quantum physicist Niels Bohr (1885–1962) said, "*Prediction is very difficult, especially if it's about the future.*" He was likely referring to a Danish proverb from the 1948 time-period.

Interestingly, Mr. Bohr also said, "*Technology has advanced more in the <u>last thirty</u> years than in the previous two thousand. The <u>exponential</u> increase in advancement will only continue.*"

If this was stated in about 1960 or earlier then his perspective starts from 1930 or sooner. Despite how difficult it is to accurately forecast even ten seconds in advance for some events, predicting the future state of some technology stands a chance if we can extend relevant historical trends. Although Mr. Bohr's first quote is slightly pessimistic his second quote is optimistic and applies to extending likely exponential trends to obtain realistic forecasts. Of course, there are many reasons why relying on brute force extrapolation often fails. Bottom line, there is merit in the approach if applied sensibly.

So, the approach in this section is to identify several current technology trends and extend them 15 years into the future. Due caution is applied with acknowledgment of expected plateaus for some trends. Why 15 years? It is a reachable goal given the current momentums and advances in areas that impact video technology and infrastructure.

The Granddaddy of Exponential Laws in Electronics

The most famous exponential curve in electronics is Moore's Law. This law states "The number of transistors on an integrated circuit (IC) would double every year [starting from 1959]." It was postulated (not called Moore's Law at the time) in a paper in Electronics Magazine in 1965 by Gordon Moore of Intel (Figure 8.1). There is no unanimous consensus on who invented the IC. Certainly Jack Kilby (Texas Instruments), Robert Noyce (Intel), and Kurt Lehovec (Sprague Electric) all played vital roles and Mr. Kilby received the Nobel Prize in Physics (2000) for his work.

The first data point on the graph in Figure 8.1 was likely for Mr. Kilby's IC, a one transistor, multiple resistor, phase-shift sinusoidal oscillator. Six years later in 1965 there were 2^6 (64) transistors on an IC. Mr. Moore's great idea was to extend the data trend with the dotted line heading to 2^{16} in 1975, a growth that was a doubling of transistor density every year. This author has extended the original graph to 2016 indicating 2^{35} (about 34 billion) transistors per IC. The original growth rate has slowed down to what is now about a doubling every 2.5 years, hence the different slope of the line heading towards 2016. The Altera/Intel Stratix 10 FPGA has approximately 30 billion transistors [Ref 1] so it's in the ballpark of 2^{35}.

The fact that Moore's Law has remained true, albeit with some slope modification, for about 57 years is amazing and provides confidence in extending its predictions year after year. There are many thoughtful researchers who continue to predict the end of Moore's Law but so far it has not been deterred. There is a joke about the projected failure of the Law: The number of people predicting the end of Moore's Law doubles every two years.

Of course, someday it will end. In the insightful paper *Universal Limits on Computation* [Ref 2], the authors put a limit on the total number bits that can be processed in a theoretical computer as 1.35×10^{120}. This fantastic number in turn requires an end to Moore's Law sometime around the year 2600. This is an unreachable upper limit and reality is a date

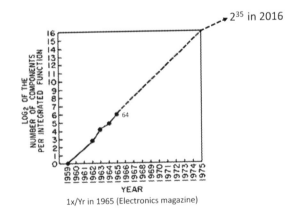

FIGURE 8.1 Moore's Law from 1959 to 2016

considerably closer, possibly near 2022 [Ref 3]. This topic is explored in more detail later in the chapter. Next, an appreciation for the power of exponential growth.

The Power of the Exponential

It is easy for humans to appreciate linear growth, but exponential thinking challenges our logic. See Figure 8.2.

If a person has a step distance of 2.5' then after 30 steps, they have progressed 75 feet. This is easy to grasp (line 'A'). On the other hand, if a person with "very long legs" makes each step double the previous then after 30 steps they have walked about 1 million miles (line 'B'). What an incredible difference with the same number of steps. This is the power of exponential growth and explains Moore's Law going from 1 to ~2^{35} transistors per-IC in just 57 years.

Note also line 'C': This shows a curve starting as exponential but slowing down and coming to an end or plateau. Adults who experienced the COVID-19 pandemic in 2020 have a very real understanding of exponential infection-growth and the meaning of "flattening the curve" as the growth tapers off. There are many examples of (sometimes fast) rising technologies sidelined due to competing choices. Instances are[1]:

- Postal/Telegraph/Phone
- Analog/Digital (SDI)
- Slide rule/Computer
- Vinyl/Tape/CD/Online
- Point-to-point/Internet
- CRT/LCD

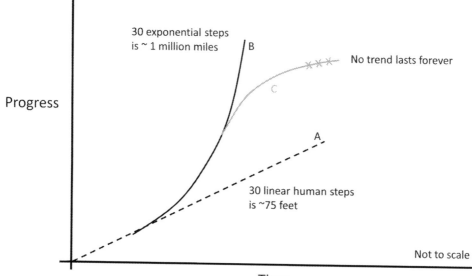

FIGURE 8.2 Linear Compared to Exponential Growth

- DVD/Netflix
- Videotape/Servers
- Custom AV/IP-COTS
- Facility data center/Cloud

A classic example is the rise and fall of the audio CD. From about 1985 to 2000 units-shipped grew about 32×, a doubling every three years on average [Ref 4]. After 2000, the CD purchase rate slowed down and is nearly dead in 2018 being replaced by online music choices. The CD example is common as technologies follow the "slow-exponential-slow" path as in 'C.' This is often called the S-curve life cycle [Ref 5]. Typically, one S-curve replaces the previous one as has occurred for the vinyl/tape/CD/online timeline. It's important to understand the concept of the S-curve when making forecasts.

By way of illustration, Figure 8.3 shows a simplified evolution of video image quality over time. The S-curves are shown "in series" but in reality, two or more will overlap as one method loses favor and another slowly replaces it. From the 'black & white' era[2] (1941) to the HDR+ era (2015 and beyond) image quality has improved many orders of magnitude following an exponential path of S-curves. The Y axis is labeled *Progress* and for the general case can represent some assigned "metric of improvement" (e.g., image resolution, bit depth, colorimetry, etc.).

With this background, let's apply these ideas to forecasting the state of video technology and infrastructure in 2036. The next sections of this chapter will consider what future growth looks like for compute, storage, networking, and related video infrastructure.

The Evolution of Infrastructure Elements

The three key elements that comprise media systems infrastructure are storage, compute, and networking (abbreviated here as S-C-N). Each will be extended with analysis on the basis for the prediction. The results developed here are applicable to any size of media system.

Storage Capacity

In this case the approach is to study the existing trends for storage and then extrapolate to about 2036. Over the period 1956–2014 the average Cumulative Annual Growth Rate

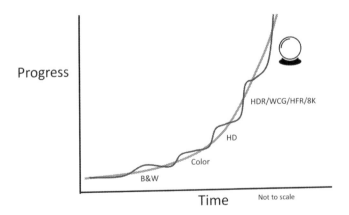

FIGURE 8.3 The Exponential and the S-Curves of Evolution for Video

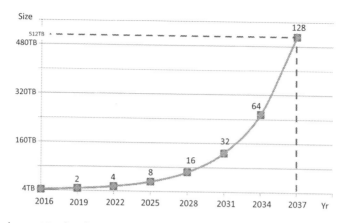

FIGURE 8.4 Predicting Hard-Disk Storage Capacity

(CAGR) for Hard Disk Drive (HDD) areal bit density was 41% and nearly identical to the Moore's Law rate for semiconductor devices over the same period [Ref 6].

For this analysis, a "conservative" future CAGR of 26% (about 2× every 3 years) is used to predict disk capacity in 2036. Figure 8.4 shows this growth starting from 4TB, a common capacity in 2016. In 2037 the estimated hard-disk capacity is about 512TB. Of course, this may seem unlikely but only if you bet against exponential growth. On the other hand, it can be argued that a constant 26% CAGR will not last for 20 years. Nonetheless, new storage methods are being invented now that could possibly fill the gap if one occurs. Remember the S-curve. In 2020, 14 TB drives are commercially available.

As the physical universe goes, so goes the digital universe; it is expanding rapidly. In 2020 it contains nearly as many digital bits as there are stars in the universe. It is doubling in size every two years, and by the end of 2020 the digital universe of data we generate will reach 44 zettabytes, or 44 trillion gigabytes [paraphrased from Ref 7]. Some of this is temporary data and will be erased after its lifecycle ends. With the death of videotape, most professional video is now stored on hard disk or LTO (or similar) tape for archive. So even with the projected large growth of storage capacity, it seems there will be ample data to fill the devices.

Compute Performance

Next, the future state of compute power, measured by the number of transistors per IC, is evaluated. There are other metrics but this one will be used given its time-honored status. Naturally, following Moore's Law is a good guide. The current growth rate is closer to 2× every 2.5 years, according to Intel CEO Brian Krzanich [Ref 8]. This is considerably less than the 2× every year that the Law started with in 1965. For our analysis we will use 2× every 3 years as with the storage case. This also applies to the Graphics Processing Unit (GPU), very useful for video processing and machine learning computation.

Modern CPUs have multiple compute cores per IC and transistor count is divided among the cores. For our purposes, the Intel 24-core Xeon Broadwell-E5 (22 active to increase yield) will be the starting benchmark with about 7.2 billion transistors. Applying a CAGR of 26%, Figure 8.5 predicts the number of cores per IC in 2037. The Y axis unit is "CoreX" and is the multiplier for the base benchmark.

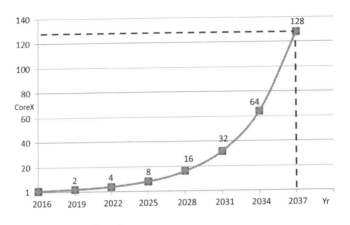

FIGURE 8.5 Predicting Compute Core Increase

With a CoreX value of 128 in 2037, this implies that the future CPU will have 3,072 cores (maybe 10% or more inactive due to chip yield) with about 922 billion transistors. As pointed out earlier and in Ref [3], it is unlikely that Moore's Law will continue unabated until 2036 even at the slower 2× per 3 years. However, the same reference also describes research to replace the traditional CMOS transistor and keep the Law alive. One way to increase the core count is to connect "chiplets" each with say N cores. A chiplet is an IC block that has been specifically designed to integrate with other similar chiplets to form larger more complex chips. So, K chiplets (N cores each) interconnected will yield K*N useful compute cores.

Thus, Moore's Law may indeed slow down considerably or even abruptly stop within the next 6–7 years. Or, it may not. Humanity has an insatiable appetite for compute power and researchers are doing their best to ensure that performance increases year after year.

Even if the transistor as we know it comes to a dead end, no doubt something will replace it as a "binary switch," the foundation of computing. For this chapter it is assumed, somehow, the Law continues even if recrafted as Version 2.0.

Software Powers Media Workflows

What about the software that executes on 3,072 cores? It is beyond the scope of this chapter to investigate the state of software in 15 years. However, there is little doubt that methods will exist to leverage this enormous compute power.

With the advent of Siri, Cortana, Google Assistant, Alexa, and others, "digital assistants" using software-based artificial intelligence (implemented using machine learning (ML) techniques) are mainstream. Many companies at NAB 2019 showcased products using ML and leveraged services from Google Cloud Platform, Amazon Web Services (AWS), IBM Waston, and Microsoft Azure Cognitive Services. The range of applications is impressive and ranges from simple metadata analysis to natural language processing (NLP) to automated program editing.

Accuracy and usefulness are improving daily. It's not too difficult to imagine AI assistants helping to run the media infrastructure of the future. Will an AI assistant ever manage and control the workflows necessary to run a media business? Consider this conclusion from a survey of 550 AI experts [Ref 9];

Results reveal a view among [550] experts that AI systems will probably (over 50%) reach overall human ability by 2040–50, and very likely (with 90% probability) by 2075.

Sometime after 2036 AI assistants may well run most operational aspects of the media enterprise. The collective prediction of experts in the field indicates a strong probability by 2075, about 40 years short of the 200-year anniversary of SMPTE.

Does this sound a little too fanciful? In his famous 1950 paper [Ref 10], Alan Turing predicted:

> I believe that in about fifty years' time [~2000] it will be possible to program computers … to play the imitation game [question-answer] so well that an average interrogator will not have more than 70 per cent chance of making the right identification after five minutes of questioning.

Basically, for a <u>machine to pass</u> the Turing test, a questioner cannot distinguish a machine from a human more than 70% of the time during a 5-min conversation. Mr. Turing's prediction time is close given the February 2011 winning display of knowledge[3] by IBM's Watson on the TV program Jeopardy [Ref 11].

Networking Capability and Performance

The final element in the S-C-N triumvirate is networking. This includes individual routers/switches and their interconnected meshes. The analysis here considers the standalone router/switch. Its capability may be measured by the *four gages* listed in Figure 8.6. Routers with hundreds of 100G and 400G ports are common for universal *leaf* and *spine* designs in 2020.

Methods to stream real-time A/V essence over Ethernet/IP are replacing the Serial Digital Interface (SDI) and AES3 (audio) methods in use today. The transition to IP is well underway utilizing the SMPTE ST2110-xx standards suite. A network's performance (lossless

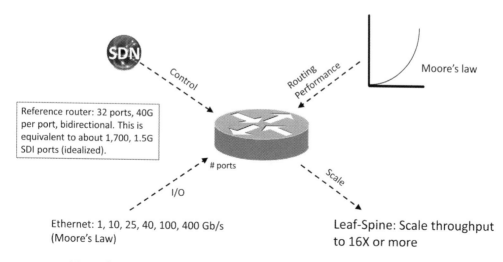

FIGURE 8.6 Networking Router Performance

transport, non-blocking switching, massive data throughput) and features will be heavily relied upon as the basis of the media facility, large and small [Ref 12].

A key gauge of "Routing Performance" is data-throughput (e.g., packets-per-second) and the ability to do deep packet inspection decision making. This metric has followed Moore's Law since the packet switching is based on silicon integrated circuits. A second gage is the number of ports and their rate (e.g., 1, 10, 25, 40, 100 Gb/s). Historically, link rates start with the IEEE standardized rate of 10 Mb/s in 1983. The 1 Tb/s link is expected in about 2025 (Ethernet Alliance). Hence, the link rate will have grown by about 100,000× in ~40 years. This is an end-to-end CAGR of about 33%, surpassing Moore's Law in a sense[4] and not as log-linear due to the uneven date spacing of each new IEEE standard.

Note that Ethernet rates are closely tied to transistor speed and not chip density. An Ethernet transceiver has very few transistors compared to say a CPU. Plus, Ethernet can use "parallel lanes" and this multiplies the native link rate by the number of lanes. With photonic wavelength division multiplexing the limit of "link" speed may reach 6.2 Tb/s [Ref 13]. Extremely high-speed Ethernet links find application in generic data trunking and other concentrated connectivity uses.

The "Scale" gauge in the figure is applied when individual routers/switches are clustered into a network. The Internet is at one end of scale. For professional media networking, the leaf-spine architecture is gaining popularity and will be a common method [Ref 14] to route AV streams in a facility much as SDI switches do today. Leaf-spine configurations enabling 1000s of HD/UHD routed video-over-IP streams are available from some vendors in 2020.

The final gauge in the figure is "Control." For most data traffic, standard routing protocols and methods suffice to steer packets across a network. However, for IP video streams in a studio/venue environment, more deterministic route control is necessary. Streams must be transported using lossless (no packets dropped) paths with guaranteed rates and very low latency end-to-end. Software-defined Networking (SDN) methods are ideal to meet the deterministic routing needs for media transport. SDN is not required for all use cases but many will benefit.

Products using SDN techniques to route IP multicast flows across switches and, in some cases, support frame accurate video switching performance [Ref 15] are common. Most modern switches support some form of SDN control, and we should expect the control dynamics to improve over time with more deep packet inspection ability. One can imagine video stream switching occurring on a frame boundary within the router itself in the future.

Using past developments in networks as a guide, expect links, meshes, and the standalone router/switch to improve markedly in the future. There is no one metric that can capture all the improvements that are likely to occur, but the performance/capability will likely increase exponentially over the next 15 years and beyond.

Infrastructure Rides the Exponential

Media systems' infrastructure is composed of many elements and the S-C-N triplet has the biggest share. As shown, these three have been riding exponential growth curves and will likely continue for a time possibly in the classic S-curve fashion. Figure 8.3 provides an example of how one S-curve is replaced by another over time.

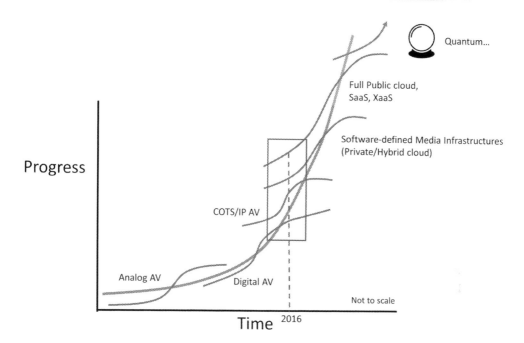

Quantum...

Full Public cloud,
SaaS, XaaS

Software-defined Media Infrastructures
(Private/Hybrid cloud)

Progress

COTS/IP AV

Analog AV

Digital AV

Not to scale

Time 2016

FIGURE 8.7 Exponential S-Curves of Media Infrastructure Progress

What might the future of the media infrastructure look like? One view is illustrated in Figure 8.7.

The Analog and Digital S-Curves

It all started in the 1940s with the advent of analog television. All production was either live or using film until the videotape recorder was invented by AMPEX in 1956. In about 1990 digital SDI transport came into being and slowly replaced analog transport. SDI has improved starting[5] at 270 Mb/s (SMPTE ST 259) and a single link data rate of at 24 Gb/s (SMPTE ST 2083-20) in 2019. This is an 88× data-rate growth in about 28 years and needed to support the ever-improving video resolutions and increasing frame rates. Video multiplexed across dual and quad links increases the aggregate transport rate at the cost of system complexity (e.g., quad-link ST 2083-22 at 96 Gb/s). However, the next S-curve labeled "COTS/IP" is gaining steam, eventually to replace SDI and other "AV custom" links and equipment.

The COTS/IP S-Curve

COTS is *commercial-off-the-shelf* IT equipment including the commodity versions of S-C-N. The intent is to use these elements to ingest, process, store, and distribute media. The IP term indicates the move towards IP/Ethernet transport to carry A/V and metadata streams. This S-curve is mature for file-based workflows and at the emergent stage for real-time IP video streaming. Both vendors and end-users appreciate the business and technical benefits of COTS/IP infrastructure compared to bespoke AV systems [Ref 16].

The SDMI S-Curve

The next S-curve taking shape is labeled Software-defined Media Infrastructure (SDMI[6]. Common themes of *software-defined* (SD) are:

- Providing "resource services" that are independent of the hardware
- Programmability of behavior
- Dynamic resource control/management

Using APIs, controllers allocate each S-C-N resource to match workloads. If 5 TB of additional storage is needed to execute a given workflow, controllers allocate it, if available, without effort from an administrator. Same for compute and networking resources in a respective way. The three models are:

- SD Compute using virtualization and containers
- SDS for storage
- SDN for networking

The full SD-based facility offers resource agility and programmable workflows. The level of user programming may become very advanced as in, "Create a new channel with these features...." SD-based infrastructure enables machine automation of workflows not possible with "just a collection of IT hardware." Of course, SD methods rely on commodity IT equipment but importantly the **software-based resource control features** give this model its benefits.

How does the SDMI differ from a generic SD-based system? Broadcast, venue events, and post media workflows require special tunings of the infrastructure that non-optimized systems may not support. Here are a few:

- Real-time media transport, point-to-multipoint streaming
- Lossless transport, very low end-to-end latency
- Precise time and AV sync support based on IEEE 1588v2 Precision Time Protocol
- Other media transport and processing features

Figure 8.8 shows an SDMI integrated with a traditional media system and any required non-IT media devices (e.g., cameras, mics, etc.). For this case the SDMI is assumed located in a private data center (e.g., broadcast operations facility). The scale of the SDMI compared to the traditional system will vary depending on many factors. Some systems will be weighted towards the traditional and other towards SDMI. Either way, traditional AV, SDMI (on premise/private), and public cloud hybrid configurations will work together in harmony.

According to DAvid® Floyer, co-founder and CTO of Wikibon, "Software-led infrastructure is a game-changer for businesses and organizations, on the same scale as the Internet was in 1995." Time will tell if this is hyperbolic speech or an accurate prediction.

Importantly, SDMI techniques can leverage the public cloud; it is inherently software-defined. Operators may have less control over some performance metrics when going public, but all file-based and many real-time AV workflows can certainly be implemented.

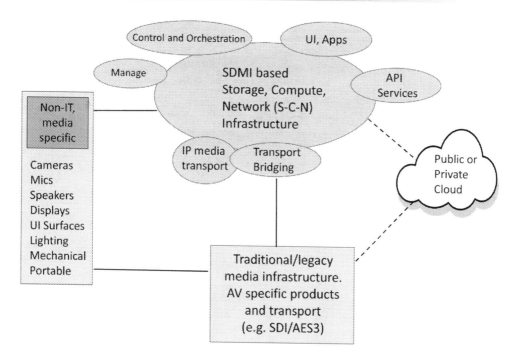

FIGURE 8.8 A Hybrid SDMI and Traditional Media System

The Public Cloud S-Curve

The next S-curve in Figure 8.7 is labeled "Full public cloud, SaaS, XaaS." This is public cloud-centric approach for executing media workflows. A few of the notable players are: Google's Cloud Platform, Amazon Web Services (AWS), and Microsoft's Azure. All services are based on mature SD principles at web-scale and services are typically sold by-the-hour.

Software-as-a-Service (SaaS, apps running in browsers) are already becoming an important aspect of many media workflows. XaaS represents "anything-as-a-Service" and implies the wide range of services from public cloud providers and their partners. Examples include Desktop-as-a-Service (DaaS), Disaster Recovery-as-a-Service (DRaaS), IDentity as a Service (IDaaS), Infrastructure-as-a-Service (IaaS), Video Encoding-as-a-Service (VEaaS), ML-as-a-service (MLaaS), and many others. These services will find use when building some media workflows. At the API level, the SMPTE is supporting "Microservices for Media" an ongoing effort to define common services for the media enterprise.

A few media vendors are building cloud-based workflow products and offer them as solutions-for-hire in some fashion. One example of this is the CLEAR software suite for broadcasters from Prime Focus Technologies [Ref 17]. This is a good example of a cloud-native media product that can work either alone or in a hybrid environment.

Expect cloud-based media solutions to mature and follow the cloud's growth patterns. It is inevitable that complete broadcast operations will be based on public clouds. The worries of security and reliability are already fading as the major cloud providers continue to show that they offer world-class security and high-availability services.

Other usage concerns are (1) access connection reliability and bandwidth and (2) limited real-time performance specs. For the first concern, access rates are ever increasing, and most cloud providers offer private 10 Gb/s (and greater) connectivity directly to their cloud, bypassing the Internet. For the second concern, there are methods to reliably transport and process real-time video despite some uncertainly on compute performance metrics.

The public cloud is only about 15 years old measured from Amazon's first IaaS service offerings in 2006. Cloud adoption is growing exponentially with Amazon's AWS reporting an average CAGR (revenue) of ~37% over the past four years. This is quite large and not representative for all cloud service suppliers. Respected research firm IDC stated in 1/2016 [Ref 18]:

> We expect worldwide spending on public cloud services will grow at a 19.4% CAGR – almost six times the rate of overall IT spending growth – from nearly $70 billion in 2015 to more than $141 billion in 2019.

From a different perspective, Statistica reports that spending on public cloud IaaS hardware and software was $25B in 2015 with forecast growth to $161B in 2023. This is a forecast CAGR of ~26% over the eight-year period [Ref 19].

The public cloud is certainly on the exponential growth part of its S-curve. However, many designers and systems integrators of media workflows are conservative and taking a wait-and-see approach. On balance, the overall benefits the cloud offers are too good to pass up despite the concerns some have. Cloud trust, security, features, reliability, performance, and ROI are all improving. Will <u>all</u> media operations be public cloud based in 2036? Highly unlikely. There will always be at least some use cases for local, on premise/truck, workflows. However, clever designers will be looking for reasons to use the cloud rather than reasons not to.

Contemporaneous S-Curves

Figure 8.7 shows four S-curves having influence during the same time period, centered around 2016. Media workflow infrastructures are being built in 2020 based on one or more of these. In time, "Digital AV" and even "COTS/IP" will become less important as "SDMI" and "Full public cloud" takes more market share over time. What is the next S-curve? Quantum entanglement for media transport? Quantum compute clouds? Time will tell.

Final Words

These predictions are a "best guess" using historical trends with nonaggressive extrapolation coupled with the thoughtful predictions of other researchers. No trend lasts forever so using S-curves provides a method to replace a dying trend with a likely startup. This approach has been successful in understanding the progress of media systems methodologies overtime (Figure 8.7). Of course, some of the conclusions in this chapter will fall short while others may surprise us and go beyond the extrapolated value. Gordon Moore in 1965 never imagined that his eponymous Law would be so long-lasting. It's not dead yet.

Albert Allen Bartlett, Professor of Physics (1923–2013) said, *"The greatest shortcoming of the human race is our inability to understand the exponential function."* This seems an exaggeration but there is truth to his words because humans have a difficult time thinking nonlinearly. If we assume that storage, compute, and networking growth metrics are all following a corresponding steep exponential then the future of the media enterprise over the next 15 years will look considerably different from that of today.

Will the conclusions presented here help you make better planning decisions? As an end-user, the discussion should improve your ability see a little farther down the road. If you are a vendor of media solutions, start thinking about "designing for the cloud." Many hardware devices need replacement in < 5 years so there will likely be 3–4 generations of on-premises equipment over 15 years; time enough for big changes. It is also true that the farther out one looks the less relevant the predictions may be. Nonetheless, it is insightful to consider the future of the media enterprise but doing so with the proverbial pinch of salt.

Source Note: The materials for this chapter are based on a talk given by the author at the SMPTE Annual Technical Conference, October 2016.

THE FUTURE IS REAL

Stan Moote, CTO – IABM

Back in 2008 I was encouraging the broadcast industry to consider cloud activities. This was a huge struggle. Being one who has always embraced technology and new business tactics, I grappled with trying to understand the reluctance to use any sort of cloud activity for broadcast. My gut feel was broadcasters didn't want to lose control. Control of what, I thought. Was it simple, plain control over every aspect of their day-to-day activities? After all, broadcasting is a unique animal, not only filled with niche, specialized equipment, but also with special, distinctive workflows. On top of these special workflows, it seemed there were always multiple exceptions too. For example, an extra pre-roll for the evening news – but only for a specific time zone; a local cut-in perhaps only once or twice a year; or even something as simple as a customized voice over to meet some regional legal requirement.

These "specials" created a whole mixture of homegrown along with custom "products" often only used in a single facility. Just one, two, or perhaps three technical types would create a company to support a facility filled with "specials." Chief Engineers had their way to practically do anything they desired to keep operations happy – often with little discussion of "Why are we really doing this?"; "Are there other options?". It wasn't the Chief Engineer's place to question, only to dream up ways to serve operations. Sometimes this even involved improving upon the concept with yet more complexities, yielding the need for historical knowledge to keep the plant running.

In came the concept of central-casting. After all, why should each station have its own master-control with operators? Now that was more than terrifying – how could that possibly work under so many special conditions? Quickly the last-mile conductivity became the "excuse" not to central-cast. This was because the larger playout centers could get the programs to the downtown core switching facility, but no further without bearing horrendous costs to make those last mile connections. The second excuse was localized ads and bumpers. How could the station create local material and use it for spot playouts? Local news/sports/weather was easy to handle, as it would typically be for a full hour, so the central-casting systems simply were not used during this time. The ad salespeople screamed – "How can I sell a last-minute ad, get it to the central-casting facility and change the rundown in minutes?"

Basically, central-casting was shelved in the late 1990s, yet a few brave souls practically (and secretly) figured it out. It wasn't so much about having local insertion; it was more about understanding how to get content and control of the content back to the central-casting operation in a timely manner. The point was to maintain control and most operational people could not accept this could be possible.

Almost invisibly, central-casting started to creep in, mainly due to the cost savings that it could provide. Along with these directives to save operating costs and improve the bottom line, local productions started to disappear. Not only was it cheaper to "buy-in" programming, the change in technology to high definition meant new capital equipment retrofits to shoot and produce local shows. Sure, this was fine for the cash cows of local news and sports, but not for full studio productions.

As I was travelling around promoting the switch to digital, it quickly became apparent to me that pretty much everywhere worldwide facilities were playing reruns of just about every syndicated program voice dubbed and subtitled for the local market. They clearly were suffering from a shortage of programs. I could see them buying in third and fourth tier programs from various countries just to fill out their playlist.

I thought, why couldn't all these empty local studios around the world start shooting productions again? Well, to do this, the economics of the local markets couldn't support this enough, yet everyone wanted local programming – they just wouldn't pay enough for it through local advertising. They needed to have a broader audience. It was also doubtful that similar local market viewers would be interested in watching programs from other markets, so it was clear that to be successful, these needed to be sold throughout the world. With their skinny margins, third party distribution consolidators could not be used. They needed to sell direct, similar to how public and state broadcasters distribute their programs.

I was still tackling why broadcasters were so strongly resisting cloud technology. Each one seemed to have different concerns; however, it all came down to fear. Not fear of the unknown, it was fear of something they couldn't touch, feel, or for that matter put their hands around: fear of the intangible. That's right, an apprehension of putting all their eggs in one basket, having no idea where that basket actually was situated and to top it off, being run by a bunch of IT nerds that had no idea or concepts of the broadcast business. The truth be-known, these IT nerds understand about having five 9s of reliability, probably in much more detail than broadcasters. They simply have a different definition of down time and maintenance schedules because they don't work in a "live" environment. For movies and very short videos, playout from the cloud was very marginal due to the lack of Internet speeds, which in turn led to buffering. This was accepted as the norm, however certainly not for television. Broadcasters were blinded by the mere fact if they started working in the cloud, they would be riding the technology curve. All would become adequate for many of their needs. They would be ahead of the competition. But how to convince them?

Taking the fact that the world needed more programs and the fear about cloud, I came up with a concept called Cloud Broadcasting in 2010. Simply put, any local program could be put up into the cloud as a file and sold to as many world markets as possible. Files didn't care about live or buffering concerns when being transferred. A cloud service simply needed to be reliable, cater worldwide, and be cost effective.

Rights management is always an issue; however, given that the programs were original local productions, rights issues were a no-brainer. So, at this point in time, focus on selling programs as files; live streaming would come later. Simply use the cloud as a tool to house and sell programs worldwide. Why not? This has been working well for e-Books, and audio books too!

Corporate LANs were no longer special, just the norm and typically running as private clouds to house storage of data files. Similar to central-casting, cloud and data centers started to creep into our industry. The mere fact of people seeing the term Cloud and Broadcasting together meant it had to be possible, without the risk of fear. This is one of the cool bits about many people in our industry: tying a few key words together unquestionably conjures up different ideas from people working in different areas of the workflow chain. Operations thought – perhaps cloud is practical for me. Chief Engineers started to tinker with other concepts like playout and IP backhauls. Moore's Law was on everyone's side. Yes, some early adopters had issues, but they learned and moved on beyond file based to actual live streams – and today even with HDR 4K!

So how does this relate to workflows in the future? The assumption is the younger generation has this all figured out, so I started to question various people square in the middle of the production process to get some insights of what they were thinking the future would hold. This was a real eye-opening experience for me, being the kind of person who embraces technology and figures out how to take advantage of it for the good of the industry. More often than not this "younger generation" could not see beyond their current workflows and job. Let me give you an example. I asked a 20 something color corrector how he saw the future about his job and work environment. He was adamant nothing would change. He would come in to work, have a bunch of files and do his job, same equipment, same workflow, same outputs. I explained to him how color correction has changed over the decades, gone from expensive suites into a single computer. He was not even aware of these changes. He never saw changes. He is trapped in what I call a technology warp. Think about it. His smart phone gets replaced bi-annually because it is cool. Not because it is faster, provides new services, strictly because it is cooler, looks hot, and perhaps has a better camera with multiple lenses and a second one for selfies. Did his workflow change on the smartphone – for sure it did. But he just accepts this, doesn't even think about the differences – that is his norm.

The more seasoned people whine about new features, reminisce about the past, and appreciate what the future will hold. Think about the future of acquisition. Cameras are everywhere. Cheap 4K cameras proliferate outside of our industry. Just like cloud, they were not mature, had cheap lenses and were pooh-poohed as being no use. Take any reality program today: they use dozens of them. Sure, they may not genlock, have proper timecode, but they do work to capture content. And this they do in such a unique, awesome way that workflows and production techniques quickly changed to accommodate the shortfalls of these cameras. Cheap meant economical, inexpensive, and cost-effective and by no means crap or unreliable, low quality.

As for DSLRs, future-wise I don't see them cutting it anymore. They are clumsy and not ergonomically friendly for video shoots. Sure, for some fixed tripod work they are great, but that's about it. Workflows don't mix with audio and DSLRs. Beyond large productions, it is also about being able to do as much as you can at the time of shooting. This may even mean that audio mixers and selectors will be built right into the cameras. I believe the key to the future is to do as much work up front as possible.

Lens sets are clearly constantly getting lighter and smaller, which may appear to contradict my previous statement. This keeps a good balance between the cheap 4K devices and professional cameras that maintain a practical form factor for seasoned camera people to use.

Beyond the constant improvement in lenses, stabilization, and sensors, I do see two evolving technologies propagating into cameras. These are Blockchain and Machine Learning.

Beyond currency, blockchain within our industry is considered as a way to firm up distribution rights, payments, and contracts. The way it does this is with multiple digital ledgers that can't be corrupted. Why would we want this with acquisition? Simple – everyone is struggling with the reality of fake news. Suppose during a shoot the camera logged into the blockchain ledger system with data such as GPS coordinates, check codes on the essence – perhaps even enough metadata to be used by recognition engines for people, locations, and objects. The essence would not be in the blockchain; however, enough defining information would be there to prove where and when it was shot. When the question comes up about this being real or not, going back to the blockchain ledgers with strict metadata controls will give proof of the pertinent details. The same goes for audio given the march of newer technologies that can create and alter words, sentences, and lips with the intent to mislead the public.

There was a lot of hesitation about SaaS when it comes to production. Companies that rely on remote editors and creatives have been well into the cloud now. The confusing part is there have been stacks and stacks of drives on editors' desks until we hit the ultimate pivotal change within the whole industry – COVID-19. Suddenly cloud was recognized as not only a completely acceptable alternative to on-site activities, but in many cases as a preferred approach.

The industry quickly learned that in the future, it is a must to have direct uploading of all proxies from the camera into the cloud while shooting, following on with the complete footage – perhaps with some mezzanine compression into the cloud allowing instant access for all parts of the workflow. As for raw sensor data footage, this will still be sometime before this is practical to use in all but higher-end productions. Again, with the appropriate smarts within the camera, raw sensor data won't be necessary for many parts of the workflow.

Machine learning fits in two completely separate paths within acquisition. The first is strictly to assist with shooting. We can program a camera to make preset zooms, etc., but it will become far more intelligent. Think of this: the camera begins to learn by the talent's facial expressions and voice subtexts when to zoom in or out, have soft focus and for that matter pan. It can even be taught that for that specific talent, there are rules set out by the producer – like always have sharp focus on a facial detail like a dark mole or chin dimples. The camera through learning would be able to look after depth of field automatically. Some would think this takes away from the creative gift camera people have. I maintain this frees them up to be creative in ways they have never considered before.

Additionally, we already are seeing how machine learning (ML) and artificial intelligence (AI) are being used to search through archives and help in finding the appropriate clips from both current and past shoots. Why not take this to the next level? Consider if the camera can be taught these search criteria. This could aid directly to help assure the correct details; scene understanding – emotions and situation – are shot, saving on reshoots and searching through endless amounts of footage.

With AI and ML being used in the production process and the availability of seeming boundless image resolutions, we can have cameras everywhere. So many cameras, it will tend to line up with the fact we have microphones everywhere – so why not cameras everywhere too? To understand the future of imagery I always look at the past and current of audio. This goes for video too. Radio had solid-state cart machines: when technology caught up, we had still video stores and clip servers. Sporting venues have microphones everywhere; now we have point-of-view (POV) cameras everywhere. This will go well beyond special events into day-to-day shoots.

This brings us back to the expensive loop discussed previously in this book, "we will fix it in post." I certainly learned the better thought out a shoot, the more cost effective in both dollars and time it was to complete the production. By having dozens of cameras, the concept of "we will fix it in post" turns into "we will create it in post." This may be true – however I do caution this could turn into people having a "remix" mentality in an attempt to make a profit rather than be creative. Hmmm – the director's cut, the producer's cut, each actor's cut, perhaps even the chief grip's cut with different camera angles? There are so many moving parts, components, and workflows during a full production to consider this as being "simplified," but are there? The Internet of things (IoT) will come to the rescue. With everything being completely connected the camera can command the lights to change to achieve the proper depth of field while maintaining the appropriate shutter to match the producer's dream. For large studio productions, green screens will be pretty much completely taken over by massive LED screens. LED sets will also provide a good percentage of the set's lighting. This again ties directly into AI/ML techniques being sympathetic with all the requirements for the cameras on the set.

The real question is, will systems be taught through AI and ML to grab content as it is being shot, taking it through the complete production process including important look and feel like color correction/grading to generate the complete scene on the spot? Perhaps even adding in compositing and creating appropriate effects too. No doubt it will happen as this is all supported by cloud stuff too.

Our format wars are no longer 4:3 or 16:9, and in the future not about 1080 vs 4K vs 8K. It will become more of a question how immersive the experience is. Immersive is typically thought of as virtual reality (VR) with goggles or in a theater; it is classically all about sound. Well, VR may be thought of as creating the ultimate immersive experience, but the potential to stimulate people's senses and draw them deep into the media production will become a key differentiator from production to production. Immersion may be as simple as sitting close to a 70" plus television. This leads us again into multiple production formats. Can you letter box immersion? Of course not; however, you can take an immersive production that is meant for sitting close to your large screen TV into a non-immersive experience. This conjures up concepts of rather than having safe action area markings in the viewfinder, rather there is a non-immersive safe area marking. It becomes a concern about pan and scan too. With sound down-mixing? Or not? This is exactly where AI and EDIDs fit into our future world.

EDID (Extended Display Identification Data) was developed in the mid-1990s to let the graphic card know the details about the monitor connected such as pixel details, sync levels, pedestal, gamma, scan rates, etc. In the web world, servers know your browser details and the web designer takes the assumption that they are in control of every pixel you see. In our world just as a streaming engine knows the type of mobile or handheld device it is streaming to, AI could take this to the next level knowing the viewing/sound devices and even the viewer's likes, dislikes, and preferences too!

I dare say some of this is happening at some level today. It will become more and more prevalent in the near future. This is all about us learning how to take advantage of the new tools available, to learn, and create new skills while tying this back to our roots of producing and distributing great content.

Getting back to cloud workflows and direct uploads from the camera, well-tuned business and production workflows can start to use content the second it is shot. Social media and web posts will build excitement about a project even during the shoot under strict controls of the content owners with assistance from blockchain. Near live interviews can be shot, approved, and posted as just part of the standard workflow process. Being in the cloud, this eliminates the geographical and country boundaries allowing the workflow and posts to be tailored towards the local market in a flash.

This may all sound simplistic; the stumbling block won't be connection bandwidth, it will be security. We will suffer a wave of cyber panics, so simply assume this and prepare for it. Don't assume only hackers are looking to steal your content. They will always find the weakest link to meet their own personal goals. The secret will be to get the mind-set changed in the future that security is everyone's responsibility. We will lose some speed and flexibility in the short term. In the long term it will become part of our daily routine and we will not even notice. The question isn't: Can you prepare your organization for this? The question is: When can you prepare your organization and workflows for this eventuality? Cyber-security has to become second nature for all involved in the process. Organizations need to teach AI machines to look for the huge blips we typically have in our network behaviors that are more event driven (like breaking news) rather than business driven (like quarter end activities).

Within Telco environments, self-healing networks have been the norm. The same goes for many public cloud operations. Until COVID-19 hit, disaster recovery (DR) was all about second sites for playout, storage, etc. – lessening the operational and profit hit due to earthquakes, hurricanes, and major network/power outages. Future-wise a big effort will go less into having redundant chains, and more about using AI/ML to sense faults and self-heal over dispersed geographies – all without human touch. Low-Earth orbit (LEO) satellites will support the control data required switch and self-heal operations without the need for reliable network connections.

At the beginning of this section I was droning on about the hundreds of customized workflows within our industry. The trend was to attempt to limit customization as it is expensive. I call this an attempt because the IABM research is showing a growth in BIY (build it yourself) solutions by end users. We will see less and less unique operational workflows and the balance between linear and on-demand continuous to settle down. The trend will be to keep adding unique natures and options into the creative process. This is exactly how tools like IMF will take on a stronger role in the future with various enhancements thanks to AI. There is a constant drive to harmonize shooting and production to deliver content in different formats (HD-UHD, SDR-HDR, WGC, Object-based audio, etc.) and via different delivery methods. These cost saving drivers will be the enabler for customizable multi-dimensional "SMART-Creative" productions. Just like how color correctors changed post – expect new tools to make a revolutionary change to the business.

As our world becomes more connected, keeping track of which time zones your coworkers are in will be more and more critical for productivity. Technology can't put us all on the same time zone; however, it certainly can assist with it. Let's accept the inevitable of the future, keep our minds, schedules apps, and timecode too set to UTC (Coordinated Universal Time) just like the Internet and air traffic controllers do. If an on-location shoot is out by 16 hours, prepare your personal environment, sleep, and meals to line up with the talent's time zone. Keep in mind two points: our bodies need "solar energy" too, so schedule this accordingly, and select a single UTC time to jam-sync all your timecode operations every day (include leap second changes as the same time – currently about every 18 months).

Will we as an industry become more risk adverse? Absolutely! Learning how to quickly parse out fads versus successful ventures, accepting and spinning up new workflows/business models due to pandemics, artistic, social and economic adaptations, as well as understanding, which are the right technologies to match your goals and aspirations, needs be assisted by a culture transformation with everyone in the industry. We can't simply be given new tools to use and accept this as the norm; everyone needs to jump in to generate new ideas, strategies, and creative techniques just like the industry pioneers have done as unmistakably explained in this book!

NOTES

1. The two examples in red text are transitions in progress.
2. Not discussed here is the "mechanical age" of television beginning in earnest in 1924–1926 with J.L. Baird and C.F. Jenkins public demonstrations of capture, transmission, and display of moving images.
3. While Watson did not officially pass the Turing Test, it showed remarkable progress towards doing so.
4. Using Moore's Law to evaluate Ethernet rate growth should be applied loosely since transistor count is not the metric but rather speed.
5. Other rates are supported in ST 259 including 360, 143, and 177 Mbit/s but 270 is the most common in use.
6. Some of the material in this section is based on the article "Software-defined Media Infrastructures", Al Kovalick, SMPTE Journal, Centennial Issue 6, August 2016.

REFERENCES

1. *Altera's 30 Billion Transistor FPGA*, http://www.gazettabyte.com/home/2015/6/28/alteras-30-billion-transistor-fpga.html.
2. *Universal Limits on Computation*, L. M. Krauss & G. D. Starkman, May 10, 2004, https://arxiv.org/pdf/astro-ph/0404510v2.pdf.
3. *After Moore's Law*, The Economist, March 12, 2016, http://www.economist.com/technology-quarterly/2016-03-12/after-moores-law.
4. *Blame It On The CD*, Jacob's Media Strategies, March 9, 2016, http://jacobsmedia.com/blame-it-on-the-cd/.
5. *Technology Life Cycle*, Wikipedia, https://en.wikipedia.org/wiki/Technology_life_cycle.
6. *HDD Areal Density Reaches 1 terabit/sq-in*, Computer History Museum, http://www.computerhistory.org/storageengine/hdd-areal-density-reaches-1-terabit-sq-in.
7. *The Digital Universe of Opportunities*, IDC, April 2014, http://www.emc.com/leadership/digital-universe/2014iview/executive-summary.htm.
8. *Intel Chief Raises Doubt over Moore's Law*, Financial Times, July 15, 2015, https://www.ft.com/content/36b722bc-2b49-11e5-8613-e7aedbb7bdb7.
9. *Future Progress in Artificial Intelligence: A Survey of Expert Opinion*, Vincent C. Müller & Nick Bostrom, Oxford University, http://www.nickbostrom.com/papers/survey.pdf.
10. *Computing Machinery and Intelligence*, A. M. Turing, Mind, Vol. 59, No. 236 (October 1950), pp. 433–460.

11. *Computer Wins on 'Jeopardy!': Trivial, It's Not*, J. Markoff, February 16, 2011, New York Times.

12. *Design Elements for Core IP Media Infrastructure*, Al Kovalick, SMPTE Motion Imaging Journal, Vol. 125, No. 2 (2016), pp. 16–23.

13. *Ethernet Roadmap 2015*, Ethernet Alliance, http://www.ethernetalliance.org/wp-content/uploads/2015/03/Front-of-Map-04-28-15.jpg.

14. *Journey of 9's – High Availability for IP Based Production Systems*, Pradeep Kathail, Charles Meyer, SMPTE Annual Technical Conference and Exhibition, SMPTE 2015, DOI: 10.5594/M001634.

15. *Software-Defined Networking*, http://en.wikipedia.org/wiki/Software-defined_networking.

16. *Video Systems in an IT Environment (2nd Ed., 2009)*, Al Kovalick, Focal Press.

17. http://www.primefocustechnologies.com/sites/default/files/files-uploded/PFT_White-paper_TVTechnology.pdf.

18. *Worldwide Public Cloud Services Spending Forecast to Double by 2019, According to IDC*, https://www.idc.com/getdoc.jsp?containerId=prUS40960516.

19. *Public Cloud Infrastructure as a Service (IaaS) Hardware and Software Spending*, Statistica, 2016, https://www.statista.com/statistics/507952/worldwide-public-cloud-infrastructure-hardware-and-software-spending-by-segment/.

Glossary of Terms

Section Editors: Chris Lennon & Clyde Smith

Media Workflow is an area fraught with a veritable alphabet soup of acronyms and terms. A glossary will be provided, helping the user to "speak the language" that is media workflow.

AQC: Automated Quality Control. This is typically carried out by a machine by providing it with specific test plans for testing specific fields (both metadata and data) and the objective is either a pass/fail or ensuring that the data is within a pre-established tolerance.

Artistic Intent: The program maker's interpretation of the images, primarily conveyed through the use of color, low-lights, and mid-tones.

ASC: The American Society of Cinematographers.

ASC CDL: The ASC Color Decision List (ASC CDL) is a format for the exchange of basic primary color grading information between equipment and software from different manufacturers. The format defines the math for three functions: Slope (Gain), Offset (lift), and Power (Gamma).

Aspect Ratio: The aspect ratio of an image describes the proportional relationship between its width and its height. It is commonly expressed as two numbers separated by a colon, as in *16:9*.

Background Plates: Background "plates" are simply shots of the background without the subject. They're called plates, as a reference to the early days of motion pictures, where backgrounds used to be printed on plates and set behind the subject.

Bayer Light Filter: A color filter array (CFA) for arranging RGB color filters on a square grid.

BNC Connector: Short for "Bayonet Neill–Concelman," it is a miniature quick connect/disconnect radio frequency connector used for coaxial cable.

BPM: Business process management is a discipline in operations management in which people use various methods to discover, model, analyze, measure, improve, optimize, and automate business processes.

CCD Image Sensor: A charge-coupled device – an integrated circuit containing an array of linked or coupled capacitors.

CCTV: Closed-circuit television (CCTV) is the use of video cameras to transmit a signal to a specific place, on a limited set of monitors.

CCU: The camera control unit (CCU) is typically part of a live television broadcast chain. It is responsible for powering the professional video camera, handling signals sent over the camera cable to and from the camera, and can be used to control various camera parameters remotely.

CGI: Computer-Generated Imagery, or special visual effects created using computer software.

CMOS Image Sensor: A complementary metal oxide semiconductor; although CCD image sensors dominated at one point, CMOS sensors have since surpassed them in usage.

CMYK: Cyan, Magenta, Yellow, and Black.

COTS: Commercial off-the-shelf, products that are commercially available and can be bought "as is."

Dailies: Unedited camera footage collected during the making of a production, also referred to as "rushes" in Canada.

DAM: Digital asset management (DAM) is a system that stores, shares, and organizes digital assets in a central location.

DBS: Direct Broadcast Satellite. Satellite television systems in which the subscribers, or end users, receive signals directly from geostationary satellites.

DCI: Digital Cinema Initiatives. Digital Cinema Initiatives, LLC (DCI) was created in March, 2002, and is a joint venture of Disney, Fox, Paramount, Sony Pictures Entertainment, Universal and Warner Bros. Studios. DCI's primary purpose is to establish and document voluntary specifications for an open architecture for digital cinema that ensures a uniform and high level of technical performance, reliability, and quality control.

Debayering: An algorithm or digital image process that is used to reconstruct a full color image from the incomplete color samples output from an image sensor overlaid with a color filter array (CFA). Most modern digital cameras acquire images using a single image sensor overlaid with a CFA, so debayering is part of the processing pipeline required to render these images into a viewable format. Many modern digital cameras can save images in a raw format allowing the user to debayer them using software, rather than using the camera's built-in firmware.

DI: Digital Intermediate is a motion picture finishing process which classically involves digitizing a motion picture and manipulating the color and other image characteristics.

Display-referred conversion: Conversion that maintains the artistic intent of the original content when converted to the new format, by calculating the light reproduced by the signal on a reference display.

DNXHD: A video codec intended to be usable as both an intermediate format suitable for use while editing and as a presentation format. DNxHD data is typically stored in an MXF container, although it can also be stored in a QuickTime container.

DOI: A digital object identifier (DOI) is a persistent identifier or handle used to identify objects. It requires an additional layer of administration for defining DOI as a URN namespace.

Dolby Atmos™: Dolby Atmos is the name of a surround sound technology by Dolby Laboratories that debuted in June 2012. Dolby Atmos technology allows up to 128 audio tracks plus associated spatial audio description metadata (most notably, location or pan automation data) to be distributed to theaters for optimal, dynamic rendering to loudspeakers based on the theater capabilities. Each audio track can be assigned to an audio channel, the traditional format for distribution, or to an audio "object."

Down Mix: In sound recording and reproduction, audio mixing is the process of combining multitrack recordings into a final mono, stereo, or surround sound product. These tracks that are blended together are done so by using various processes such as equalization and compression. Reproducing 5.1 channels of audio through a lesser number of speakers requires a process called downmixing. Simply put, downmixing combines the Left, Right, Center, Left surround, and Right surround channels in a somewhat logical manner to drive stereo or mono speakers.

DSLR: A digital camera that combines the optics and the mechanisms of a single-lens reflex camera with a digital imaging sensor.

DTH: Direct To Home television is a method of receiving satellite television by means of signals transmitted from direct-broadcast satellites.

DTS-X: In early 2015 the company DTS launched DTS:X, its competitor to Dolby Atmos and a similar surround sound technology. Rather than define a fixed number of channels, one for each speaker, DTS-X allows the "location" (direction from the listener) of "objects" (audio tracks) to be specified as polar coordinates. The audio processor is then responsible for dynamically rendering sound output depending on the number and position of speakers available.

DVB: Digital Video Broadcasting (DVB) is a set of international open standards for digital television. DVB standards are maintained by the DVB Project, an international industry consortium, and are published by a Joint Technical Committee (JTC) of the European Telecommunications Standards Institute (ETSI), European Committee for Electrotechnical Standardization (CENELEC) and European Broadcasting Union (EBU).

DVS: Descriptive Video Service, also referred to as Audio description, video description, described video, or more precisely called a visual description, is a form of narration used to provide information surrounding key visual elements in a media work (such as a film or television program, or theatrical performance) for the benefit of blind and visually impaired consumers.

EDL: An edit decision list is used in the post-production process of film editing and video editing. The list contains an ordered list of reel and timecode data representing where each video clip can be obtained in order to conform the final cut.

EFP: Electronic Field Production, which is a video production which takes place in the field, outside of a formal television studio, in a practical location or special venue.

EIDR: Entertainment Identifier Registry is a global unique identifier system for a broad array of audiovisual objects, including motion pictures, television, and radio programs.

ENG: Electronic news gathering (ENG) is when reporters and editors make use of electronic video and audio technologies in order to gather and present news.

EOM: Start of Material/Media. The timecode at which a piece of content ends.

Flash Memory: A non-volatile memory chip used for storage and for transferring data between a personal computer (PC) and digital devices.

Foley: Common everyday sounds like footsteps, clothes rustling, and door creaks are normally re-recorded. These sounds are re-created and recorded by a Foley Artist. A Foley Editor takes the new sounds and synchronizes them to the edit of the content.

Friends and Family screening: Test screenings have always been an important part of a film's post-production process. These screening to friendly members (Friends and Family) are used to gauge audience responses to important story beats and serve as a means of fine tuning the motion picture.

Full or Narrow Video Range: Describes the code range available for the video signal. For mastering purposes, the full range of video is utilized, while for transmission and distribution the narrow range is utilized, thus permitting control information to be embedded in the video signal. Recommendation ITU-R BT.2100 specifies two different signal representations: "narrow" and "full." Television can only be narrow range – full range signals will simply be clipped. More information can be found in the following document – SMPTE RP 2077.

Genlock: A device for maintaining synchronization between two different video signals, or between a video signal and a computer or audio signal, enabling video images and computer graphics to be mixed.

Grassmann's Laws: These describe empirical results about how the perception of mixtures of colored lights (i.e., lights that co-stimulate the same area on the retina) composed of different spectral power distributions can be algebraically related to one another in a color matching context. Discovered by Hermann Grassmann these "laws" are actually principles used to predict color match responses to a good approximation under photopic and mesopic vision.

Hard Disk Drive (HDD): An electro-mechanical data storage device that uses magnetic storage to store and retrieve digital data using one or more rigid rapidly rotating platters coated with magnetic material.

HDRI: High-dynamic-range imaging is a technique used in photographic imaging and films, and in ray-traced computer-generated imaging, to reproduce a greater range of luminosity than what is possible with standard digital imaging or photographic techniques.

HDSLR: Hybrid Digital Single Lens Reflex. A camera that can take both stills and videos.

HDTV: A television system providing an image resolution of substantially higher resolution than standard definition television (SDTV).

HFC: Hybrid fiber-coaxial is a telecommunications industry term for a broadband network that combines optical fiber and coaxial cable. It has been commonly employed globally by cable television operators since the early 1990s.

High Dynamic Range: High Dynamic Range (HDR) refers to the extended (or high) contrast range that is the displayed currently in either Theatrical Dolby Vision (EDR) or high contrast and high brightness television, approximating a range of 10,000:1 or greater.

HLG: Hybrid Log-Gamma (HLG) is a backwards-compatible high dynamic range (HDR) standard that was jointly developed by the BBC and NHK.

IMF: Interoperable Master Format (IMF) is a SMPTE standard for providing a single, interchangeable master file format and structure for the distribution of content between businesses around the world. IMF provides a framework for creating a true file-based final master.

IRD: An integrated receiver/decoder is an electronic device used to pick up a radio-frequency signal and convert digital information transmitted in it.

ITU: The International Telecommunication Union is a specialized agency of the United Nations that is responsible for issues that concern information and communication technologies. The ITU establishes standards and best practices for the broadcast industry.

JSON: JavaScript Object Notation is an open standard file format, and data interchange format, that uses human-readable text to store and transmit data objects consisting of attribute–value pairs and array data types (or any other serializable value).

Locked Picture: This is a stage in editing a film or editing a television production and establishes the point where all changes to the film or television program cut have been done and approved.

Long Form Content: Typically considered any content exceeding two minutes in length. Normally, this implies program content.

LTO: Linear Tape-Open (LTO) is a magnetic tape data storage technology originally developed in the late 1990s as an open standards alternative to the proprietary magnetic tape formats that were available at the time.

MAM: A Media Asset Management system provides a single repository for storing and managing video and massive multi-media files.

M&E: M&E is an abbreviation for Music and Effects. Generally denoting an audio track consisting of the mixed and combined music and effects stems of a film, an M&E track should not contain any intelligible on-screen dialogue. An M&E track consisting only of the combined music and effects stems of a film will be missing all the production effects and sounds that were inherent in the dialogue stem and is considered "unfilled." This is not suitable for foreign dubbing. When production effects and ambience from a film's dialogue stem are edited into the M&E track or recreated through editorial, it is considered "filled," and may be suitable for foreign dubbing.

Metadata: "Data about data," generally data that describes other data.

MSO: A multiple-system operator (MSO) is an operator of multiple cable or direct-broadcast satellite television systems.

MTF: Modulation Transfer Function is the equivalent of OTF in most cases, but neglects phase effects.

MUSE: Multiple sub-Nyquist sampling encoding was an analog high-definition television standard, using dot-interlacing and digital video compression to deliver 1125-line (1920×1035) high definition video signals to the home.

NCS: The Natural Color System (NCS) is a proprietary perceptual color model. It is based on the color opponency hypothesis of color vision, first proposed by German physiologist Ewald Hering.

Nearfield Mix: NearField Mix (aka Home Theater master) is a modified mix that optimizes the sound mix for the home environment. Typically, this modifies the center dialogue channel, limits the dynamic range and modifies the bass to conform to a narrower home screen.

Neutral/Balance Grade: This is normally carried out as a precursor to the color grading for the motion picture. The idea here is to balance/neutralize the shots to fix any glaring issues with regard to white balance, exposure, contrast, or any other technical parameter. In this way, all the shots are in balanced, and from here the creative process of setting specific looks that enhance the story can begin.

NTSC: National Television Standards Committee, named for the group that originally developed the black & white and subsequently color television system that is used in the United States, Japan, and many other countries.

Nyquist Frequency: Is half of the sampling rate of a discrete signal processing system. It is named after electronic engineer Harry Nyquist. When the function domain is time, sample rates are usually expressed in samples per second, and the unit of Nyquist frequency is cycles per second (Hertz).

Optical Disk: A computer disk that uses optical storage techniques and technology to read and write data.

OTF: Optical Transfer Function, which specifies how different spatial frequencies are handled by the system.

OTT: An Over-the-Top (OTT) media service is a streaming media service offered directly to viewers via the Internet. OTT bypasses cable, broadcast, and satellite television platforms, the companies that traditionally act as a controller or distributor of such content.

PCR: Production Control Room, sometimes also known as a Studio Control Room (SCR), or the Gallery, is where the composition of a program takes place.

PQ: Stands for 'Perceptual Quantizer' which refers to a new 'gamma' or transfer function for HDR. It uses the human eye as the perceptual basis for its signal-to-light relationship.

Preview Screening: Similar to Friends and Family screening above.

ProRes: Developed by Apple, ProRes is a line of intermediate codecs which means they are intended for use during video editing, and not for practical end-user viewing. This is achieved by only using intra-frame compression, where each frame is stored independently and can be decoded with no dependencies on other frames.

PSE: Photosensitive Epilepsy (PSE) is a form of epilepsy in which seizures are triggered by visual stimuli that form patterns in time or space, such as flashing lights; bold, regular patterns; or regular moving patterns.

PTZ: A pan–tilt–zoom camera (PTZ camera) is a camera that is capable of remote directional and zoom control.

Python: Python is an interpreted, high-level, general-purpose programming language. Created by Guido van Rossum and first released in 1991, Python's design philosophy emphasizes code readability with its notable use of significant whitespace.

QC: Quality Control. The process of verifying that the final delivery properly correlates to the creative direction that was provided for generating the content. All Audio, Video, and metadata aspects are subject to rigorous QC – both automated and manual.

REST: Representational state transfer (REST) is a software architectural style that defines a set of constraints to be used for creating Web services. Web services that conform to the REST architectural style, called RESTful Web services, provide interoperability between computer systems on the Internet. RESTful Web services allow the requesting systems to access and manipulate textual representations of Web resources by using a uniform and predefined set of stateless operations. Other kinds of Web services, such as SOAP Web services, expose their own arbitrary sets of operations.

RGB: Red, Green, Blue. An additive color model in which red, green, and blue light are added together in various ways to reproduce a broad array of colors.

R-Y (and B-Y): These represent color difference signals.

Scene-referred conversion: Conversion usually used on Live or multi-camera studio or outside broadcast programs where matching will change the appearance of content after conversion. It is used to color match cameras and graphics in live production, by calculating the light falling on the camera sensor.

SDI: Serial digital interface (SDI) is a family of digital video interfaces first standardized by SMPTE.

SDTV: Standard Definition Television. The two common SDTV signal types are 576i, with 576 interlaced lines of resolution, derived from the European-developed PAL and SECAM systems, and 480i based on the American NTSC system.

SMPTE: The Society of Motion Picture and Television Engineers, founded in 1916, is a global professional association of engineers, technologists, and executives working in the media and entertainment industry. SMPTE establishes standards and best practices for the motion picture and television industry.

SOA: Service-oriented architecture is a style of software design where services are provided to the other components by application components, through a communication protocol over a network.

SOAP: Simple Object Access Protocol is a messaging protocol specification for exchanging structured information in the implementation of web services in computer networks.

SOM: Start of Material/Media. The timecode at which a piece of content begins.

Standard Dynamic Range: Standard Dynamic Range (SDR) refers to the contrast range that is displayed currently in theatrical or television displays. In Theatrical exhibition, the Standard Dynamic Range (SDR) is normally 2000:1, or about ten stops. For SDR television in the home, with a nominal peak at 100 cd/m^2, the dynamic range is about 128:1 or about seven stops.

Storyboard: A storyboard is a graphic organizer in the form of illustrations or images displayed in sequence for the purpose of pre-visualizing a motion picture, animation, motion graphic, or interactive media sequence.

Telecine: Telecine is the process of transferring motion picture film into video and is performed in a color suite.

The Look: The intended appearance of colors and tones within a scene.

Tintype: A photograph made by creating a direct positive on a thin sheet of metal coated with a dark lacquer or enamel and used as the support for photographic emulsion.

Tone-mapping: Converting content that has a certain range (in f-stops or contrast range) to another range. This is commonly done when mapping from one display device (say High Dynamic Range) to another device (say Standard Dynamic Range).

Triax: Also known as triaxial cables, it is a type of electrical cable similar to coaxial cable, but with the addition of an extra layer of insulation and a second conducting sheath. It provides greater bandwidth and rejection of interference than coax, but is more expensive.

TTML: Timed Text Markup Language (TTML), previously referred to as Distribution Format Exchange Profile (DFXP), is an XML-based W3C standard for timed text in online media and was designed to be used for the purpose of authoring, transcoding, or exchanging timed text information presently in use primarily for subtitling and captioning functions.

UDDI: Universal Description, Discovery, and Integration is a platform-independent, Extensible Markup Language protocol that includes a (XML-based) registry by which businesses worldwide can list themselves on the Internet, and a mechanism to register and locate web service applications.

UHDTV: Ultra High Definition Television, which encompasses 4K and 8K resolutions, and is considered the successor to HDTV.

UUID: A universally unique identifier (UUID) is a 128-bit number used to identify information in computer systems.

VCR: A video cassette recorder (VCR) is a tape recorder designed to record and play back video and audio material on a magnetic tape videocassette.

VFX: Visual Effects. Visual Effects is the process by which imagery is created or manipulated outside the context of a live action shot in film making.

VFX Pulls: The process whereby Visual Effects shots are "pulled" from the live action shots – as either reference plates or as background plates for effects work. Normally, the Visual Effects Editor will request the specific shots to be pulled from the Raw Camera footage and have these delivered to the Visual Effects company to begin their work.

VTR: A video tape recorder (VTR) is a tape recorder designed to record and play back video and audio material on magnetic tape.

WSDL: The Web Services Description Language is an XML-based interface description language that is used for describing the functionality offered by a web service.

XML: eXtensible Markup Language is a markup language that defines a set of rules for encoding documents in a format that is both human-readable and machine-readable.

YAML: An acronym for "YAML Ain't Markup Language" is a human-readable data-serialization language. It is commonly used for configuration files and in applications where data is being stored or transmitted. YAML targets many of the same communications applications as Extensible Markup Language (XML) but has a minimal syntax which intentionally differs from SGML. It uses both Python-style indentation to indicate nesting, and a more compact format that uses [...] for lists and {...} for maps making YAML 1.2 a superset of JSON.

YIQ: The color space used by the NTSC color TV system, employed mainly in North and Central America, and Japan. I stands for in-phase, while Q stands for quadrature, referring to the components used in quadrature amplitude modulation.

Index

Printed in the United States
By Bookmasters